McGraw-H

500
Linear Algebra
Questions

Also in McGraw-Hill's 500 Questions Series

McGraw-Hill's

500
Linear Algebra
Questions

Ace Your College Exams

Seymour Lipschutz, PhD

New York Chicago San Francisco Lisbon London Madrid Mexico City
Milan New Delhi San Juan Seoul Singapore Sydney Toronto

1 2 3 4 5 6 7 8 9 10   QFR/QFR   1 9 8 7 6 5 4 3 2

QA
1845
.L57
2013

ISBN   978-0-07-179799-3
MHID   0-07-179799-8

e-ISBN   978-0-07-179800-6
e-MHID   0-07-179800-5

Library of Congress Control Number 2011944583

McGraw-Hill products are available at special quantity discounts to use as premiums and sales promotions or for use in corporate training programs. To contact a representative, please e-mail us at bulksales@mcgraw-hill.com.

This book is printed on acid-free paper.

CONTENTS

INTRODUCTION

You've taken a big step toward success in linear algebra by purchasing *McGraw-Hill's 500 Linear Algebra Questions*. We are here to help you take the next step and score high on your first-year exams!

This book gives you 500 exam-style questions that cover all the most essential course material. Each question is clearly explained in the answer key. The questions will give you valuable independent practice to supplement your regular textbook and the ground you have already covered in your class.

This book and the others in the series were written by experienced teachers who know the subject inside and out and can indentify crucial information as well as the kinds of questions that are most likely to appear on exams.

You might be the kind of student who needs to study extra before the exam for a final review. Or you might be the kind of student who puts off preparing until the last minute before the test. No matter what your preparation style, you will benefit from reviewing these 500 questions, which closely parallel the content and degree of difficulty of the questions on actual exams. These questions and the explanations in the answer key are the ideal last-minute study tool.

If you practice with all the questions and answers in this book, we are certain you will build the skills and confidence needed to excel on your exams.

—Editors of McGraw-Hill Education

CHAPTER 1

Vectors

Vectors in $\mathbf{R}^n$, Vector Addition, Scalar Multiplication

Definition: A *vector* $\mathbf{u}$ in $\mathbf{R}^n$ is a list of n real numbers: $\mathbf{u} = [u_1, u_2, \ldots, u_n]$. For vectors $\mathbf{u} = [u_1, u_2, \ldots, u_n]$ and $\mathbf{v} = [v_1, v_2, \ldots, v_n]$, the *sum* $\mathbf{u} + \mathbf{v}$ and the *scalar product* $k\mathbf{u}$ of $\mathbf{u}$ are defined as follows: $\mathbf{u} + \mathbf{v} = [u_1 + v_1, u_2 + v_2, \ldots, u_n + v_n]$ and $k\mathbf{u} = [ku_1, ku_2, \ldots, ku_n]$.

A *column vector* is a vector written vertically rather than horizontally.

1. (A) For vectors $\mathbf{u}$ and $\mathbf{v}$, when is $\mathbf{u} = \mathbf{v}$?
   (B) Which of the following vectors are equal?

   $\mathbf{u}_1 = [1, 2, 3], \mathbf{u}_2 = [2, 3, 1], \mathbf{u}_3 = [1, 3, 2], \mathbf{u}_4 = [2, 3, 1]$

2. Find:
   (A) $[3, -4, 5, -6] + [1, 1, -2, 4]$
   (B) $[1, 2, -3] + [4, -5]$
   (C) $-3[4, -5, -6]$
   (D) $-[6, 7, 8]$

3. Compute:

   (A) $\begin{bmatrix} 7 \\ -4 \\ 2 \end{bmatrix} + \begin{bmatrix} -3 \\ -1 \\ 5 \end{bmatrix}$

   (B) $5\begin{bmatrix} -2 \\ 3 \\ 4 \end{bmatrix}$

4. Let $\mathbf{u} = [2, -7, 1]$, $\mathbf{v} = [-3, 0, 4]$, $\mathbf{w} = [0, 5, -8]$. Find:
   (A) $3\mathbf{u} - 4\mathbf{v}$
   (B) $2\mathbf{u} + 3\mathbf{v} - 5\mathbf{w}$

5. Write $\mathbf{w} = [1, 9]$ as a linear combination of the vectors $\mathbf{u} = [1, 2]$ and $\mathbf{v} = [3, -1]$.

6. Write $\mathbf{v} = [2, -3, 4]$ as a linear combination of the vectors $\mathbf{u}_1 = [1, 1, 1]$, $\mathbf{u}_2 = [1, 1, 0]$, $\mathbf{u}_3 = [1, 0, 0]$.

Summation Symbol

7. Let $f(k)$ be an algebraic expression involving an integer variable k.
   (A) For $n \geq 1$, define the expression $S_n = \sum_{k=1}^{n} f(k)$.
   (B) For $n_1 \leq n_2$, define $\sum_{k=n_1}^{n_2} f(k)$.

8. Compute:

   (A) $\sum_{k=1}^{4} k^3$

   (B) $\sum_{j=2}^{5} j^2$

Dot (Inner) Product and Norm

Definition: Let $\mathbf{u} = [u_1, u_2, \ldots, u_n]$ and $\mathbf{v} = [v_1, v_2, \ldots, v_n)$ be vectors in $\mathbf{R}^n$. The *dot* (or *inner* or *scalar*) *product* of $\mathbf{u}$ and $\mathbf{v}$, denoted by $\mathbf{u} \cdot \mathbf{v}$, is defined to be the scalar obtained by multiplying corresponding components and adding the resulting products; that is,

$$\mathbf{u} \cdot \mathbf{v} = u_1 v_1 + u_2 v_2 + \cdots + u_n v_n = \sum_{k=1}^{n} u_k v_k .$$

The *norm* or *length* of $\mathbf{u}$, denoted by $\|\mathbf{u}\|$, is the nonnegative square root of $\mathbf{u} \cdot \mathbf{u}$, that is,

$$\| \mathbf{u} \| = \sqrt{\mathbf{u} \cdot \mathbf{u}} = \sqrt{u_1^2 + u_2^2 + \cdots + u_n^2} .$$

A vector $\mathbf{u}$ is a *unit vector* if $\|\mathbf{u}\| = 1$ (or, equivalently, if $\mathbf{u} \cdot \mathbf{u} = 1$).

9. Find $\mathbf{u} \cdot \mathbf{v}$ where:

   (A) $\mathbf{u} = [2, -3, 6]$, $\mathbf{v} = [8, 2, -3]$
   (B) $\mathbf{u} = [3, -5, 2, 1]$, $\mathbf{v} = [4, 1, -2, 5]$
   (C) $\mathbf{u} = [1, -2, 3, -4]$, $\mathbf{v} = [6, 7, 1, -2]$

10. Find:

    (A) $\|\mathbf{u}\|$ if $\mathbf{u} = [3, -12, 4]$
    (B) $\|\mathbf{v}\|$ if $\mathbf{v} = [2, -3, 8, -5]$
    (C) $\|\mathbf{w}\|$ if $\mathbf{w} = [-3, 1, -2, 4, -5]$

11. For $\mathbf{u} = [1, k, -2, 5]$, find k such that $\|\mathbf{u}\| = \sqrt{39}$.

12. Let $\mathbf{v}$ be a nonzero vector. Show that the vector $\hat{\mathbf{v}}$, defined by

$$\hat{\mathbf{v}} = \frac{1}{\|\mathbf{v}\|}\mathbf{v} = \frac{\mathbf{v}}{\|\mathbf{v}\|},$$

is a unit vector in the same direction as $\mathbf{v}$. (The process of finding $\hat{\mathbf{v}}$ is called *normalizing* $\mathbf{v}$.)

13. Normalize

    (A) $\mathbf{v} = (12, -3, -4)$
    (B) $\mathbf{w} = (4, -2, -3, 8)$

14. Normalize $\mathbf{v} = \left(\frac{1}{2}, \frac{2}{3}, -\frac{1}{4}\right)$.

Theorem 1.1 (Cauchy–Schwarz inequality): $|\mathbf{u} \cdot \mathbf{v}| \le \|\mathbf{u}\|\,\|\mathbf{v}\|$

15. For $\mathbf{u}$ and $\mathbf{v}$ in $\mathbf{R}^n$, prove Minkowski's inequality that $\|\mathbf{u} + \mathbf{v}\| \le \|\mathbf{u}\| + \|\mathbf{v}\|$ using theorem 1.1. *Cauchy–Schwarz inequality:*

$$lu * vl \le \|\mathbf{u}\|\,\|\mathbf{v}\| \tag{1.1}$$

Distance, Angles, Projections

Definition: The *distance* between vectors $\mathbf{u}$ and $\mathbf{v}$ in $\mathbf{R}^n$, denoted by $d(\mathbf{u}, \mathbf{v})$, is defined as $d(\mathbf{u}, \mathbf{v}) = \|\mathbf{u} - \mathbf{v}\|$. The *angle* θ between vectors $\mathbf{u}$ and $\mathbf{v}$ in $\mathbf{R}^n$ is defined by $\cos\theta = (\mathbf{u} \cdot \mathbf{v})/(\|\mathbf{u}\|\,\|\mathbf{v}\|)$. For vectors $\mathbf{u}$ and $\mathbf{v} \ne \mathbf{0}$ in $\mathbf{R}^n$, the *projection* of $\mathbf{u}$ onto $\mathbf{v}$ is the vector denoted and defined by $\text{proj}(\mathbf{u}, \mathbf{v}) = [(\mathbf{u} \cdot \mathbf{v})/\|\mathbf{v}\|^2]\mathbf{v}$.

16. Find $d(\mathbf{u}, \mathbf{v})$ where:
    (A) $\mathbf{u} = (1, 7)$, $\mathbf{v} = (6, -5)$
    (B) $\mathbf{u} = (3, -5, 4)$, $\mathbf{v} = (6, 2, -1)$

17. For $\mathbf{u} = (2, k, 1, -4)$ and $\mathbf{v} = (3, -1, 6, -3)$, find k such that $d(\mathbf{u}, \mathbf{v}) = 6$.

18. Using the Cauchy–Schwarz inequality (1.1), show that θ is a unique real number in the interval $[0, \pi]$.

19. Find $\cos \theta$ where θ is the angle between:
    (A) $\mathbf{u} = (1, -2, 3)$ and $\mathbf{v} = (3, -5, -7)$
    (B) $\mathbf{u} = (4, -3, 1, 5)$ and $\mathbf{v} = (2, 6, -1, 4)$

20. Find $\text{proj}(\mathbf{u}, \mathbf{v})$ where:
    (A) $\mathbf{u} = (1, -2, 3)$ and $\mathbf{v} = (2, 5, 4)$
    (B) $\mathbf{u} = (4, -3, 1, 5)$ and $\mathbf{v} = (3, 6, -4, 1)$

Orthogonality

Definition: The vector $\mathbf{u}$ is said to be *orthogonal* to $\mathbf{v}$ if $\mathbf{u} \cdot \mathbf{v} = 0$.

21. Which of the vectors $\mathbf{u} = (5, 4, 1)$, $\mathbf{v} = (3, -4, 1)$, $\mathbf{w} = (1, -2, 3)$, if any, are orthogonal?

22. Find k so that $\mathbf{u} = (1, k, -3)$ and $\mathbf{v} = (2, -5, 4)$ are orthogonal.

23. If $\mathbf{u}$ and $\mathbf{v}$ are orthogonal to $\mathbf{w}$, show that $\mathbf{u} + \mathbf{v}$ and any scalar multiple $k\mathbf{u}$ are orthogonal to $\mathbf{w}$.

24. Let $\mathbf{u}_1$, $\mathbf{u}_2$, $\mathbf{u}_3$ be nonzero vectors orthogonal to each other. Let $\mathbf{w} = x\mathbf{u}_1 + y\mathbf{u}_2 + z\mathbf{u}_3$. Show that

$$x = (\mathbf{w} \cdot \mathbf{u}_1)/\|\mathbf{u}_1\|^2, \quad y = (\mathbf{w} \cdot \mathbf{u}_2)/\|\mathbf{u}_2\|^2, \quad z = (\mathbf{w} \cdot \mathbf{u}_3)/\|\mathbf{u}_3\|^2. \quad (1.2)$$

25. Show that $\mathbf{u}_1 = (1, -2, 3)$, $\mathbf{u}_2 = (1, 2, 1)$, $\mathbf{u}_3 = (-8, 2, 4)$ are orthogonal to each other.

26. Write $\mathbf{w} = (13, -4, 7)$ as a linear combination of vector $\mathbf{u}_1$, $\mathbf{u}_2$, $\mathbf{u}_3$ in Problem 25.
    [Hint: Use (1.2).]

Vectors in R³, ijk Notation, Cross Product

Definition: A *spatial vector* is a vector in $\mathbf{R}^3$. Special notation is used for such vectors:

$$\mathbf{i} = (1, 0, 0), \quad \mathbf{j} = (0, 1, 0), \quad \mathbf{k} = (0, 0, 1).$$

Then any vector $\mathbf{u} = (a, b, c)$ in $\mathbf{R}^3$ can be expressed uniquely in the form $\mathbf{u} = a\mathbf{i} + b\mathbf{j} + c\mathbf{j}$.

27. Suppose $\mathbf{u} = 3\mathbf{i} + 5\mathbf{j} + 2\mathbf{k}$ and $\mathbf{v} = 4\mathbf{i} - 8\mathbf{j} + 7\mathbf{k}$. Find:
    (A) $\mathbf{u} + \mathbf{v}$
    (B) $3\mathbf{u} - 2\mathbf{v}$
    (C) $\mathbf{u} \cdot \mathbf{v}$

28. Find the following determinants of order two where $\begin{vmatrix} a & b \\ c & d \end{vmatrix} = ad - bc$:

    (A) $\begin{vmatrix} 3 & 4 \\ 5 & 9 \end{vmatrix}$

    (B) $\begin{vmatrix} 2 & -1 \\ 4 & 3 \end{vmatrix}$

    (C) $\begin{vmatrix} 4 & 5 \\ 3 & -2 \end{vmatrix}$

29. Find the negative of the following determinants of order two:

    (A) $\begin{vmatrix} 3 & 6 \\ 4 & 2 \end{vmatrix}$

    (B) $\begin{vmatrix} 7 & -5 \\ 3 & 2 \end{vmatrix}$

    (C) $\begin{vmatrix} 4 & -1 \\ 8 & -3 \end{vmatrix}$

    (D) $\begin{vmatrix} -4 & -3 \\ 2 & -2 \end{vmatrix}$

Definition: Let $\mathbf{u} = (a_1, a_2, a_3)$ and $\mathbf{v} = (b_1, b_2, b_3)$ be vectors in $\mathbf{R}^3$. The *cross product* of $\mathbf{u}$ and $\mathbf{v}$ is denoted and defined by

$$\mathbf{u} \times \mathbf{v} = (a_2 b_3 - a_3 b_2, a_3 b_1 - a_1 b_3, a_1 b_2 - a_2 b_1).$$

The three components of $\mathbf{u} \times \mathbf{v}$ may be expressed as determinants using the following array where we put the vector $\mathbf{v}$ under the vector $\mathbf{u}$:

$$\begin{bmatrix} a_1 & a_2 & a_3 \\ b_1 & b_2 & b_3 \end{bmatrix}$$

(i) Cover the first column and take the determinant.
(ii) Cover the second column and take the negative of the determinant.
(iii) Cover the third column and take the determinant.

30. Find $\mathbf{u} \times \mathbf{v}$ where:

(A) $\mathbf{u} = (1, 2, 3), \mathbf{v} = (4, 5, 6)$
(B) $\mathbf{u} = (7, 3, 1), \mathbf{v} = (1, 1, 1)$

31. Prove $\mathbf{u} \times \mathbf{v}$ is orthogonal to both $\mathbf{u}$ and $\mathbf{v}$.

32. Find a unit vector $\mathbf{u}$ orthogonal to $\mathbf{v} = (1, 3, 4)$ and $\mathbf{w} = (2, -6, 5)$.

Complex Vectors

Definition: $\mathbf{C}$ denotes the *set of complex numbers*. Here z and w are complex numbers; a, b, x, y are real numbers; and $i = \sqrt{-1}$ (meaning $i^2 = -1$). Also $z = a + bi$ is the *standard form* of a complex number where a is called the *real part* of z, written Re z, and b is called the *imaginary part* of z, written Im z.

33. Let $z = 2 + 3i$ and $w = 5 - 2i$. Find:

(A) $z + w$
(B) $z - w$
(C) zw

34. Given $i^1 = i, i^2 = -1, i^3 = -i, i^4 = 1$, find:

(A) i^{39}
(B) i^{174}
(C) i^{252}
(D) i^{317}

Definition: The *complex conjugate* of $z = a + bi$ is defined and denoted by $\bar{z} = a - bi$.

The *absolute value* of z is denoted and defined by $|z| = \sqrt{z\bar{z}}$.

35. Find the complex conjugate of each complex number:
    (A) $6 + 4i$
    (B) $7 - 5i$
    (C) $4 + i$
    (D) $-3 - i$

36. Note that if $z = a + bi$, then $z\bar{z} = a^2 + b^2$. Find $z\bar{z}$ where:
    (A) $z = 2 - 3i$
    (B) $z = 4 + 5i$
    (C) $z = 6 - 2i$

37. Find $|z|$ where:
    (A) $z = 3 - 4i$
    (B) $z = 5 - 2i$
    (C) $z = -7 + i$
    (D) $z = -1 - 4i$

38. Express each fraction in the form $a + bi$:

    (A) $\dfrac{1}{3 - 4i}$

    (B) $\dfrac{2 - 7i}{5 + 3i}$

39. Consider $\mathbf{u} = (7 - 2i, 2 + 5i)$ and $\mathbf{v} = (1 + i, -3 - 6i)$ in $\mathbf{C}^2$. Find:
    (A) $\mathbf{u} + \mathbf{v}$
    (B) $(3 - i)\mathbf{v}$
    (C) $(1 + i)\mathbf{u} + (2 - i)\mathbf{v}$

Definition: Suppose $\mathbf{u} = (z_1, z_2, \ldots, z_n)$ and $\mathbf{v} = (w_1, w_2, \ldots, w_n)$ are vectors on $\mathbf{C}^n$. The *dot* (or *inner* or *scalar*) *product* of $\mathbf{u}$ and $\mathbf{v}$ is denoted and defined by

$$\mathbf{u} \cdot \mathbf{v} = z_1 \bar{w}_1 + z_2 \bar{w}_2 + \cdots + z_n \bar{w}_n = \sum z\bar{w},$$

and the *norm* of **u** is denoted and defined by

$$\|\mathbf{u}\| = \sqrt{\mathbf{u} \cdot \mathbf{u}} = z_1 \overline{z}_1 + z_2 \overline{z}_2 + \cdots + z_n \overline{z}_n.$$

40. Let $\mathbf{u} = (2 + 3i, 4 - 1, 2i)$ and $\mathbf{v} = (3 - 2i, 5, 4 - 6i)$ in $\mathbf{C}^3$. Find:
  (A) $\mathbf{u} \cdot \mathbf{v}$
  (B) $\|\mathbf{u}\|$
  (C) $\|\mathbf{v}\|$

CHAPTER 2

Matrices

This chapter uses capital letters A, B, C, ... to denote matrices, and lowercase letters a, b, c, x, y, ... to denote scalars. Unless otherwise specified, all scalars will be real numbers.

Matrices, Matrix Addition, Scalar Multiplication

Definition: An $m \times n$ matrix $A = [a_{ij}]$ is a rectangular array of mn scalars.

41. Find the rows, columns, and size of the matrix $A = \begin{bmatrix} 1 & 2 & 3 \\ 4 & 5 & 6 \end{bmatrix}$.

42. Given matrices A and B, when is $A = B$?

43. Find x, y, z, w if $\begin{bmatrix} x+y & 2z+w \\ x-y & z-w \end{bmatrix} = \begin{bmatrix} 3 & 5 \\ 1 & 4 \end{bmatrix}$.

Definition: Suppose $A = [a_{ij}]$ and $B = [b_{ij}]$ are matrices of the same size. Then $A + B = [a_{ij} + b_{ij}]$ and $kA = [ka_{ij}]$.

44. Find $2A - 3B$ where $A = \begin{bmatrix} 1 & -2 & 3 \\ 4 & 5 & -6 \end{bmatrix}$ and $B = \begin{bmatrix} 3 & 0 & 2 \\ -7 & 1 & 8 \end{bmatrix}$.

Matrix Multiplication

Definition: The product of a matrix with one row (row matrix) $[a_i]$ and a matrix with one column (column matrix) $[b_i]$, with the same number of elements, is their inner product:

$$[a_i][b_i] = a_1 b_1 + a_2 b_2 + \cdots + a_n b_n.$$

45. Find:

(A) $[8, -4, 5] \begin{bmatrix} 3 \\ 2 \\ -1 \end{bmatrix}$,

(B) $[6, -1, 7, 5] \begin{bmatrix} 4 \\ -9 \\ -3 \\ 2 \end{bmatrix}$.

Definition: Suppose that $A = [a_{ik}]$ is an $m \times p$ matrix and $B = [b_{kj}]$ is a $p \times n$ matrix. Then the product $AB = [c_{ij}]$ is the $m \times n$ matrix for which

$$c_{ij} = a_{i1}b_{1j} + a_{i2}b_{2j} + \cdots + a_{ip}b_{pj} = \sum_{k=1}^{p} a_{ik}b_{kj}.$$

That is, the ij-entry of AB is the product of the ith row of A and the jth column of B.

46. Find the product AB for $A = \begin{bmatrix} 1 & 3 \\ 2 & -1 \end{bmatrix}$ and $B = \begin{bmatrix} 2 & 0 & -4 \\ 3 & -2 & 6 \end{bmatrix}$.

47. Find AB where $A = \begin{bmatrix} 2 & 3 & -1 \\ 4 & -2 & 5 \end{bmatrix}$ and $B = \begin{bmatrix} 2 & -1 & 0 & 6 \\ 1 & 3 & -5 & 1 \\ 4 & 1 & -2 & 2 \end{bmatrix}$.

48. Find two matrices such that AB and BA are defined and have the same size but $AB \neq BA$.

Transpose of a Matrix

Definition: The *transpose* of a matrix A, denoted A^T, is the matrix obtained by writing the rows of A, in order, as the columns. Thus the transpose of a row vector is a column vector, and the transpose of a column vector is a row vector.

49. Find A^T for $A = \begin{bmatrix} 1 & 2 & 3 \\ 4 & -5 & -6 \end{bmatrix}$.

50. Given $A = \begin{bmatrix} 1 & 3 & 5 \\ 6 & -7 & -8 \end{bmatrix}$, find A^T and $(A^T)^T$.

51. Let $A = \begin{bmatrix} 1 & 2 \\ 3 & -4 \end{bmatrix}$ and $B = \begin{bmatrix} 5 & 0 \\ -6 & 7 \end{bmatrix}$. Find:

(A) $(AB)^T$,
(B) $A^T B^T$ [Note: $(AB)^T \neq A^T B^T$.].

Elementary Row Operations, Pivots, Row Reduction

Definition: The following are the elementary row operations.

[E_1] Interchange the ith row and the jth row: $R_i \leftrightarrow R_j$.

[E_2] Multiply the ith row by a nonzero scalar k: $R_i \rightarrow kR_i$, $k \neq 0$.

[E_3] Replace the ith row by k times the jth row plus the ith row: $R_i \rightarrow kR_j + R_i$.

52. Show that each elementary row operation has an inverse operation of the same type.

Definition: Matrix A is row equivalent to matrix B if B can be obtained from A by a sequence of elementary row operations.

53. Express the following row operation in terms of the above elementary row operations:

$$[E] \quad R_i \rightarrow k'R_j + kR_i, \ k \neq 0.$$

54. Apply each of the following operations to $A = \begin{bmatrix} 1 & 2 & 3 & 4 \\ 5 & 6 & 7 & 8 \\ 3 & -4 & 5 & -6 \end{bmatrix}$:

(A) $R_2 \leftrightarrow R_3$.

(B) $R_1 \rightarrow 3R_1$.

(C) $R_3 \rightarrow -3R_1 + R_3$.

55. Suppose that a_{ij} is a nonzero element in a matrix A. Show that each of the following row operations, which changes the kth row of A, yields a 0 in the kj-position of A:

(A) $R_k \rightarrow (-a_{kj}/a_{ij})R_i + R_k$,

(B) $R_k \rightarrow -a_{kj} R_i + a_{ij} R_k$.

(The above element a_{ij}, which is used to produce 0's above and/or below it, is called the *pivot* of the operations.)

56. Produce 0's below the pivot 2 in the matrix $\begin{bmatrix} 2 & 1 & -3 & 4 \\ 3 & 4 & 1 & -2 \\ 5 & -2 & 3 & 0 \end{bmatrix}$.

57. Find the leading nonzero entries in the following matrices:

(A) $\begin{bmatrix} 0 & 1 & -3 & 4 & 6 \\ 4 & 0 & 2 & 5 & -3 \\ 0 & 0 & 7 & -2 & 8 \end{bmatrix}$

(B) $\begin{bmatrix} 0 & 0 & 0 & 0 & 0 \\ 1 & 2 & 3 & 4 & 5 \\ 0 & 0 & 5 & -4 & 7 \end{bmatrix}$

(C) $\begin{bmatrix} 0 & 2 & 2 & 2 & 2 \\ 0 & 3 & 1 & 0 & 0 \\ 0 & 0 & 0 & 0 & 0 \end{bmatrix}$.

Definition: A matrix A is called an *echelon matrix* or is said to be in *echelon form* if
(i) any zero rows are on the bottom of the matrix, and
(ii) each leading nonzero entry is to the right of the leading nonzero entry in the preceding row.

58. Which, if any, of the matrices in question 57 are in echelon form?

Algorithm 2.1: The following algorithm row reduces an arbitrary matrix $A = [a_{ij}]$ to echelon form. (The term *row reduce* shall mean to transform A by row operations.)

Step 1. Find the first column with a nonzero entry; call it the j_1-column.

Step 2. Interchange the rows so that a nonzero entry appears in the first row of the j_1-column (that is, so that $a_{1j_1} \neq 0$).

Step 3. Use a_{1j_1} as a pivot to obtain 0s below a_{1j_1}; that is, for each $i > 1$, apply the row operation $R_i \rightarrow -a_{ij_1} R_1 + a_{1j_1} R_i$ or $R_i \rightarrow (-a_{ij_1} / a_{1j_1}) R_1 + R_i$.

Step 4. Repeat Steps 1, 2, and 3 with the submatrix formed by all the rows except the first.

Step 5. Continue the above process until the matrix is in echelon form.

59. Row reduce the following matrix A to echelon form: $A = \begin{bmatrix} 1 & 2 & -3 & 0 \\ 2 & 4 & -2 & 2 \\ 3 & 6 & -4 & 3 \end{bmatrix}$.

60. Row reduce the following matrix A to echelon form: $A = \begin{bmatrix} 0 & 1 & 3 & -2 \\ 2 & 1 & -4 & 3 \\ 2 & 3 & 2 & -1 \end{bmatrix}$.

Row Canonical Form, Gaussian Elimination

Definition: *A* matrix *A* is said to be in *row canonical form if*
 (i) *A* is an echelon matrix and each leading nonzero entry is 1, and
 (ii) each leading nonzero entry is the only nonzero entry in its column.

61. Which of the following echelon matrices, whose leading nonzero entries have been boxed, are in row canonical form?

(A)
$$\begin{bmatrix} \boxed{2} & 3 & 2 & 0 & 4 & 5 & -6 \\ 0 & 0 & \boxed{7} & 1 & -3 & 2 & 0 \\ 0 & 0 & 0 & 0 & 0 & \boxed{6} & 0 \\ 0 & 0 & 0 & 0 & 0 & 0 & 0 \end{bmatrix}$$

(B)
$$\begin{bmatrix} \boxed{1} & 2 & 3 \\ 0 & 0 & \boxed{1} \\ 0 & 0 & 0 \\ 0 & 0 & 0 \end{bmatrix}$$

(C)
$$\begin{bmatrix} 0 & \boxed{1} & 3 & 0 & 0 & 4 & 0 \\ 0 & 0 & 0 & \boxed{1} & 0 & -3 & 0 \\ 0 & 0 & 0 & 0 & \boxed{1} & 2 & 0 \\ 0 & 0 & 0 & 0 & 0 & 0 & \boxed{1} \end{bmatrix}$$

Algorithm 2.2 (Gaussian Elimination Algorithm): The following algorithm, which consists of two main steps, reduces an arbitrary matrix *A* to row canonical form.

Step 1. Use Algorithm 2.1 to reduce *A* to echelon form, denoting the leading nonzero entries by $a_{1j_1}, a_{2j_2}, \ldots, a_{rj_r}$.

Step 2. Since $a_{rj_r} \neq 0$, multiply the last nonzero row R_r by $1/a_{rj_r}$. Then use $a_{rj_r} = 1$ as a pivot to obtain 0's above the pivot. Repeat the process with $R_{r-1}, R_{r-2}, \ldots, R_2$. Finally, if necessary, multiply R_1 by $1/a_{1j_1}$ to make $a_{1j_1} = 1$.

The matrix is now in row canonical form. Step 2 is sometimes called *back-substitution* since the leading nonzero entries are used as pivots in the reverse order, from the bottom up.

Problems 62–63: Reduce each matrix to row canonical form.

62. $A = \begin{bmatrix} 2 & 3 & 4 & 5 & 6 \\ 0 & 0 & 3 & 2 & 5 \\ 0 & 0 & 0 & 0 & 4 \end{bmatrix}$.

63. $B = \begin{bmatrix} 2 & 2 & -1 & 6 & 4 \\ 4 & 4 & 1 & 10 & 13 \\ 6 & 6 & 0 & 20 & 19 \end{bmatrix}$.

64. One speaks of "an" echelon form of a matrix A, but "the" row canonical form of A. Why?

65. Let $A = [a_{ij}]$ be a square echelon matrix in triangular form (that is, all entries below the diagonal are 0), and suppose every diagonal $a_{ii} \neq 0$. Find the row canonical form of A.

66. Exhibit all the row canonical forms for 2×2 matrices.

Block Matrices

67. A matrix A may be partitioned into a system of smaller matrices, called *blocks,* by a set of horizontal and vertical lines. The matrix A is then called a *block matrix.* Give the size of each of the following block matrices (which are partitionings of the same matrix):

(A) $\left[\begin{array}{cc|cc|c} 1 & -2 & 0 & 1 & 3 \\ 2 & 3 & 5 & 7 & -2 \\ \hline 3 & 1 & 4 & 5 & 9 \end{array}\right]$

(B) $\left[\begin{array}{ccc|cc} 1 & -2 & 0 & 1 & 3 \\ \hline 2 & 3 & 5 & 7 & -2 \\ \hline 3 & 1 & 4 & 5 & 9 \end{array}\right]$

68. Suppose matrices A and B are partitioned into block matrices $A = [A_{ij}]$ and $B = [B_{ij}]$ where corresponding blocks A_{ij} and B_{ij} have the same size. Find

(A) sum $A + B$,
(B) scalar multiple kA.

69. Suppose matrices U and V are partitioned into blocks where the number of columns of each block U_{ik} is equal to the number of rows of each block V_{kj}. Find the product UV.

70. Compute AB using block multiplication where $A = \begin{bmatrix} 1 & 2 & 1 \\ 3 & 4 & 0 \\ 0 & 0 & 2 \end{bmatrix}$, $B = \begin{bmatrix} 1 & 2 & 3 & 1 \\ 4 & 5 & 6 & 1 \\ 0 & 0 & 0 & 1 \end{bmatrix}$.

CHAPTER 3

Systems of Linear Equations

Linearity, Solutions, Linear Equations in One Unknown

71. Consider an equation E in n unknowns $x_1, x_2, \ldots, x_n$.

    (A)   When is it linear?

    (B)   When is it degenerate?

72. Determine whether the following equation is linear: $3x + ky - 8z = 16$.

73. Determine if $x = 3$, $y = 2$, $z = 1$ is a solution of the (linear) equation $x + 2y - 3z = 4$.

Theorem 3.1: Consider the degenerate linear equation $0x_1 + 0x_2 + \cdots + 0x_n = b$.

  (i) If the constant $b \neq 0$, the equation has no solution.

 (ii) If $b = 0$, then every vector in $\mathbf{R}^n$ is a solution.

74. Describe the solutions of the equation:

    (A) $x + 3y + x - 3 = 2y + 2x + y$

    (B) $4y - x - 3y + 3 = 2 + x - 2x + y + 1$

Theorem 3.2: Consider the linear equation $ax = b$.

  (i) If $a \neq 0$, then $x = b/a$ is the unique solution.

 (ii) If $a = 0$ but $b \neq 0$, there is no solution.

(iii) If $a = 0$ and $b = 0$, every scalar k is a solution.

75. Solve:

    (A) $4x - 12$

    (B) $5x = 0$

    (C) $2x - 5 - x = x + 3$

    (D) $4 + x - 3 = 2x + 1 - x$

Linear Equations in Two Unknowns, One Equation in Many Unknowns

76. Let E be the equation $2x + y = 4$.
(A) Find three solutions of E.
(B) Plot the graph of E.

77. Consider a system of two linear equations, call them L_1 and L_2, in unknowns x and y. A *solution* of the system is a pair $\mathbf{u} = (k_1, k_2)$ that satisfies both equations.
(A) Describe geometrically the case where the system has a unique solution, and give an example.
(B) Describe geometrically the case where the system has no solution, and give an example.
(C) Describe geometrically the case where the system has an infinite number of solutions, and give an example.

78. Solve the system
$$L_1 : 2x + 5y = 8$$
$$L_2 : 3x - 2y = -7$$

79. Find the *leading unknown* and its position p in the equation
$0x_1 + 0x_2 + 8x_3 - 4x_4 + 0x_5 - 7x_6 = 2$.

Theorem 3.3: Consider a nondegenerate linear equation $a_1x_1 + a_2x_2 + \cdots + a_nx_n = b$, where $n > 1$; let the leading unknown be x_p.
(i) Any set of values for the unknowns x_j with $j \neq p$ will yield a unique solution of the equation. (The unknowns x_j are called *free variables*, since one can assign any values to them.)
(ii) Every solution of the equation is obtained in (i). (The set of all solutions is called the *general solution* of the equation.)

80. Consider the equation $0x + 3y - 4z = 5$, or simply $3y - 4z = 5$. Find:
(A) three particular solutions
(B) the general solution

System of *m* Equation in *n* Unknowns

Definition: A system of m equations in n unknowns has the standard form

$$L_1: \quad a_{11}x_1 + a_{12}x_2 + \cdots + a_{1n}x_n = b_1$$
$$L_2: \quad a_{21}x_1 + a_{22}x_2 + \cdots + a_{2n}x_n = b_2$$
$$\vdots$$

$$\cdots\cdots\cdots\cdots\cdots\cdots\cdots\cdots\cdots\cdots\cdots\cdots \qquad (3.1)$$

$$L_m: \quad a_{m1}x_1 + a_{m2}x_2 + \cdots + a_{mn}x_n = b_m$$

81. Find the number of unknowns in the system $x + 2z = 7$, $3x - 5y = 4$.

82. Consider the system

$$x_1 + 2x_2 - 5x_3 + 4x_4 = 3$$
$$2x_1 + 3x_2 + x_3 - 2x_4 = 1$$

Determine whether each is a solution of the system:

(A) $\mathbf{v} = (-8, 4, 1, 2)$

(B) $\mathbf{u} = (-8, 6, 1, 1)$

Remark 3.1: The following are the *elementary operations* on the system (3.1):

[E$_1$] Interchange the ith equation and the jth equation: $L_i \leftrightarrow L_j$.

[E$_2$] Multiply the ith equation by a nonzero scalar: $L_i \to kL_i$ ($k \neq 0$).

[E$_3$] Replace the ith equation by k times the jth equation plus the ith equation: $L_i \to kL_j + I_{\cdot i}$.

[E] Replace the ith equation by k' times the jth equation plus k (nonzero) times the ith equation: $L_i \to k' L_j + kL_i$ ($k \neq 0$).

83. Consider the system

$$L_1 : x - 2y + 3z = 5$$
$$L_2 : 2x + y - 4z = 1$$
$$L_3 : 3x + 2y - 7z = 3$$

Apply each operation to the system:

(A) $L_2 \leftrightarrow L_3$

(B) $L_2 \to 3L_2$

(C) $L_3 \to -3L_1 + L_3$

84. What can one say about the solution set of a system that contains the degenerate equation

(A) $0x_1 + 0x_2 + \cdots + 0x_n = b$ with $b \neq 0$,

(B) $0x_1 + 0x_2 + \cdots + 0x_n = 0$?

Systems in Triangular and Echelon Form

Definition: A system of linear equations is in *triangular form* if the number of equations is equal to the number of unknowns and if x_k is the leading unknown of the kth equation. The paradigm follows where all $a_{kk} \neq 0$:

$$a_{11}x_1 + a_{12}x_2 + \ldots \ldots \ldots \ldots + a_{1,n-1}x_{n-1} + a_{1n}x_n = b_1$$
$$a_{22}x_2 + \ldots \ldots \ldots \ldots + a_{2,n-1}x_{n-1} + a_{2n}x_n = b_2$$
$$\ldots \ldots \ldots \ldots \ldots \ldots \ldots \ldots \ldots \ldots \ldots \quad (3.2)$$
$$a_{n-1,n-1}x_{n-1} + a_{n-1,n}x_n = b_{n-1}$$
$$a_{nn}x_n = b_n$$

Algorithm 3.1 (Back-substitution algorithm): The following algorithm finds the unique solution of a triangular system.

Step 1. Solve the last equation of (3.2) for the last unknown x_n:

$$x_n = \frac{b_n}{a_{nn}}.$$

Step 2. Substitute this value for x_n in the next-to-last equation, and solve for the next-to-last unknown x_{n-1}:

$$x_{n-1} = \frac{b_{n-1} - a_{n-1,n}(b_n/a_{nn})}{a_{n-1,n-1}}$$

General Step: Determine x_k by substituting the previously obtained values in the kth equation:

$$x_k = \frac{b_k - \sum\limits_{m-k+1}^{n} a_{km}x_m}{a_{kk}}.$$

The process ceases when we determine the first unknown x_1.

The solution is unique because, at each step of the algorithm, the value of x_k is uniquely determined.

85. Solve the system

$$2x - 3y + 5z - 2t = 9$$
$$5y - z + 3t = 1$$
$$7z - t = 3$$
$$2t = 8.$$

Definition: A system of linear equations is in *echelon form* if no equation is degenerate and if the leading unknown in each equation is to the right of the leading unknown of the preceding equation. The paradigm follows:

$$a_{11}x_1 + a_{12}x_2 + \cdots + a_{1n}x_n = b_1$$
$$a_{2j_2}x_{j_2} + a_{2 \cdot j_2 + 1}x_{j_2 + 1} + \cdots + a_{2n}x_n = b_2$$
$$\cdots\cdots\cdots\cdots\cdots\cdots\cdots\cdots\cdots\cdots\cdots\cdots\cdots\cdots$$
$$a_{rj_r}x_{j_r} + a_{r, j_r + 1}x_{jr+1} + \cdots + a_{rn}x_n = b_r$$

(3.3)

where $1 < j_2 < \cdots < j_r$ and where $a_{11} \neq 0$, $a_{2j_2} \neq 0$, ..., $a_{rj_r} \neq 0$. Note $r \leq n$.

86. In an echelon form, any unknown that is not a leading unknown is termed a *free variable*. Determine the free variables in the system

$$3x + 2y - 5z - 6s + 2t = 4$$
$$z + 8s - 3t = 6$$
$$s - 5t = 5.$$

Theorem 3.4: The system (3.3) of linear equations in echelon form has a unique solution if $r = n$ and has one solution for each specification of the $n - r$ free variables if $r < n$.

87. Find three particular solutions of the system

$$x + 4y - 3z + 2t = 5$$
$$z - 4t = 2.$$

88. Find the general solution to the system in Question 87:
(A) in the free-variable form
(B) in parametric form

Gaussian Elimination

Algorithm 3.2 (Gaussian Elimination Algorithm): The following algorithm reduces the general system (3.1) to echelon (possibly triangular) form or determines that the system has no solution.

Step 1. Interchange equations so that the first unknown, x_1, appears with a nonzero coefficient in the first equation; i.e., arrange that $a_{11} \neq 0$.

Step 2. Use a_{11} as a *pivot* to eliminate x_1 from all the equations except the first equation. That is, for each $i > 1$, apply the elementary operation

$$[E_3]: \quad L_i \rightarrow -(a_{i1}/a_{11})L_1 + L_i \quad \text{or} \quad [E]: \quad L_i \rightarrow -a_{i1}L_1 + a_{11}L_1.$$

Step 3. Examine each new equation L to see if it is degenerate:

(a) If L has the form $0x_1 + 0x_2 + \cdots + 0x_n = 0$, then delete L from the system.

(b) If L has the form $0x_1 + 0x_2 + \cdots + 0x_n = b \neq 0$, then exit from the algorithm, as the system has no solution.

Step 4. Repeat Steps 1, 2, and 3 with the subsystem formed by all the equations, excluding the first equation.

Step 5. Continue the above process until the system is in echelon form or a degenerate equation is obtained in Step 3(b).

89. Show that Step 3(a) in the Gaussian algorithm may be replaced by Step 3(a′).

Step 3(a′). If L has the form $0x_1 + 0x_2 + \cdots + 0x_n = 0$ or if L is a multiple of another equation, then delete L from the system.

90. Solve each system.

(A) $2x + y - 2z = 10$
$3x + 2y + 2z = 1$
$5x + 4y + 3z = 4$

(B) $x + 2y - 3z = 1$
$2x + 5y - 8z = 4$
$3x + 8y - 13z = 7$

(C) $x + 2y - 2z = -1$
$3x - y + 2z = 7$
$5x + 3y - 4z = 2$

91. Solve the system

$$2x - 5y + 3z - 4s + 2t = 4$$
$$3x - 7y + 2z - 5s + 4t = 9$$
$$5x - 10y - 5z - 4s - 7t = 22.$$

Theorem 3.5: Any system of linear equations has either

(i) a unique solution,

(ii) no solution, or

(iii) an infinite number of solutions.

92. Determine the values of k so that the following system has:
  (A) a unique solution
  (B) no solution
  (C) an infinite number of solutions

$$x - 2y = 1$$
$$x - y + kz = -2$$
$$ky + 4z = 6$$

Systems of Linear Equations in Matrix Form

93. Use a matrix product to represent the general system (3.1) and find its *augmented matrix*.

94. Rewrite the following system as a matrix equation:

$$2x + 3y - 4z = 7$$
$$x - 2y - 5z = 3.$$

95. Solve each system using its augmented matrix.
  (A) $x - 2y - 3z = 4$
     $2x - 3y + z = 5$
  (B) $x + y - 2z + 4t = 5$
     $2x + 2y - 3z + t = 3$
     $3x + 3y - 4z - 2t = 1$
  (C) $x + 2y - 3z - 2s + 4t = 1$
     $2x + 5y - 8z - s + 6t = 4$
     $x + 4y - 7z + 5s + 2t = 8$

96. The rank of a matrix A, written rank(A), is the maximum number of linearly independent variables in A. How is rank(A) related to any echelon form of A?

Theorem 3.6: A system of linear equations $AX = B$ has a solution if and only if the rank of the coefficient matrix A is equal to the rank of the augmented matrix $[A, B]$.

Homogeneous Systems

Definition: A system of linear equations is *homogeneous* if all constant terms are zero or, in matrix form, if $AX = 0$.

Theorem 3.7: Suppose the echelon form of a homogeneous system $AX = 0$ has s free variables. Let $u_1, u_2, \ldots, u_s$ be the solutions obtained by setting one of the free variables equal to one and the remaining free variables equal to zero. Then $u_1, u_2, \ldots, u_s$ form a *basis* for the solution space W of $AX = 0$. [This means that any solution of the system can be expressed as a *unique* linear combination of $u_1, u_2, \ldots, u_s$; furthermore, the *dimension* of W is dim $(W) = s$.]

97. Find the dimension and a basis for the solution space W of each homogeneous system.

   (A) $x + 3y - 2z + 5s - 3t = 0$
   $2x + 7y - 3z + 7s - 5t = 0$
   $3x + 11y - 4z + 10s - 9t = 0$

   (B) $x + 2y - 3z + 2s - 4t = 0$
   $2x + 4y - 5z + s - 6t = 0$
   $5x + 10y - 13z + 4s - 16t = 0$

98. Find the general solution of the homogeneous system in Problem 97(a).

Theorem 3.8: A homogeneous system of linear equations with more unknowns than equations has a nonzero solution.

99. Determine, without solving, whether the following system has a nonzero solution:

$$x_1 - 2x_2 + 3x_3 - 2x_4 = 0$$
$$3x_1 - 7x_2 + 5x_3 + 4x_4 = 0$$
$$4x_1 + 3x_2 + 5x_3 - 22x_4 = 0.$$

Systems of Linear Equations as Vector Equations

Remark 3.2: The general system (3.1) can be replaced by a single vector equation

$$x_1 \begin{bmatrix} a_{11} \\ a_{21} \\ \vdots \\ a_{m1} \end{bmatrix} + x_2 \begin{bmatrix} a_{12} \\ a_{22} \\ \vdots \\ a_{m2} \end{bmatrix} + \cdots + x_n \begin{bmatrix} a_{1n} \\ a_{2n} \\ \vdots \\ a_{mn} \end{bmatrix} = \begin{bmatrix} b_1 \\ b_2 \\ \vdots \\ b_m \end{bmatrix}$$

or, if $\mathbf{u}_1$, $\mathbf{u}_2$, ..., $\mathbf{u}_n$ and $\mathbf{v}$ denote the (column) vectors,

$$x_1\mathbf{u}_1 + x_2\mathbf{u}_2 + \cdots + x_n\mathbf{u}_n = \mathbf{v}.$$

Thus $\mathbf{v}$ is a linear combination of $\mathbf{u}_1$, $\mathbf{u}_2$, ..., $\mathbf{u}_n$ if and only if the system has a solution.

100. (A) Convert the following vector equation to an equivalent system of linear equations and solve:

$$\begin{bmatrix} 1 \\ -6 \\ 5 \end{bmatrix} = x \begin{bmatrix} 1 \\ 2 \\ 3 \end{bmatrix} + y \begin{bmatrix} 2 \\ 5 \\ 8 \end{bmatrix} + z \begin{bmatrix} 3 \\ 2 \\ 3 \end{bmatrix}.$$

(B) Write $\mathbf{v} = (1, -2, 5)$ as a linear combination of $\mathbf{u}_1 = (1, 1, 1)$, $\mathbf{u}_2 = (1, 2, 3)$, $\mathbf{u}_3 = (2, -1, 1)$.

CHAPTER 4

Square Matrices

A matrix A is *square* if it has the same number of rows as columns.

Diagonal, Trace, Identity Matrix, Commuting Matrices

Definition: The *diagonal* (*or main diagonal*) of an n-square matrix $A = [a_{ij}]$ consists of the elements $a_{11}, a_{22}, \ldots, a_{nn}$, and the *trace* of A, written tr(A), is the sum of the diagonal elements.

101. Find the diagonal and trace of each matrix:

(A) $A = \begin{bmatrix} 1 & 2 & 3 \\ 4 & 5 & 6 \\ 7 & 8 & 9 \end{bmatrix}$

(B) $B = \begin{bmatrix} t-2 & 3 \\ -4 & t+5 \end{bmatrix}$

(C) $C = \begin{bmatrix} 1 & 2 & -3 \\ 4 & -5 & 6 \end{bmatrix}$

Definition: The n-square *identity* (or *unit*) matrix, denoted by I_n, or simply I, is the n-square matrix with 1's on the diagonal and 0's elsewhere.

102. (A) Find the trace of the identity matrices $I_2 = \begin{bmatrix} 1 & 0 \\ 0 & 1 \end{bmatrix}$ and $I_3 = \begin{bmatrix} 1 & 0 & 0 \\ 0 & 1 & 0 \\ 0 & 0 & 1 \end{bmatrix}$

(B) Find the trace of I_n.

103. The Kronecker delta function δ_{ij} is defined by $\delta_{ij} = \begin{cases} 0 & \text{if } i \neq j \\ 1 & \text{if } i = j \end{cases}$. Thus $I = [\delta_{ij}]$. Use this to show that, for any $m \times n$ matrix A,

(A) $I_m A = A$,

(B) $A I_n = A$.

Definition: For any scalar k, the *scalar matrix* D_k is the matrix kI (that is, ks on the diagonal and 0s elsewhere). A square matrix $D = [d_{ij}]$ is *diagonal* if its nondiagonal entries are all zero; it is sometimes denoted by $D = \text{diag}(d_{11}, d_{22}, \ldots, d_{nn})$.

104. Find the scalar matrices of orders 2, 3, and 4 corresponding to $k = 5$.

105. Write out the matrices $\text{diag}(3, -7, 2)$, $\text{diag}(4, -5)$, $\text{diag}(6, -3, -9, -1)$.

106. Matrices A and B *commute* if $AB = BA$. Show that $A = \begin{bmatrix} 1 & 2 \\ 3 & 4 \end{bmatrix}$ and $B = \begin{bmatrix} 5 & 4 \\ 6 & 11 \end{bmatrix}$ commute.

Powers of Matrices

Definition: The *nonnegative integral powers* of a square matrix M may be defined recursively by

$$M^0 = I, \ M^1 = M, \text{ and } M^{r+1} = MM^r.$$

If $f(x) = \sum a_i x^i$, then $f(M) = \sum a_i M^i$.

107. Let $A = \begin{bmatrix} 1 & 2 \\ 4 & -3 \end{bmatrix}$.

  (A) Find A^2.
  (B) Find A^3.

108. For matrix A in Problem 107, find $f(A)$ for the polynomial $2x^3 - 4x + 5$.

109. For matrix A in Problem 107, show that A is a zero of the polynomial $g(x) = x^2 + 2x - 11$.

Invertible Matrices, Inverses

Definition: A square matrix A is *invertible* if there exists a (square) matrix B such that $AB = BA = I$, where I is the identity matrix. The matrix B is called the *inverse* of A.

Theorem 4.1: $AB = I$ if and only if $BA = I$.

(Hence we need only show that $AB = I$ to show that A and B are inverses.)

110. Show that $A = \begin{bmatrix} 1 & 0 & 2 \\ 2 & -1 & 3 \\ 4 & 1 & 8 \end{bmatrix}$ and $B = \begin{bmatrix} -11 & 2 & 2 \\ -4 & 0 & 1 \\ 6 & -1 & -1 \end{bmatrix}$ are inverses.

Remark 4.1: The inverse of the matrix $A = \begin{bmatrix} a & b \\ c & d \end{bmatrix}$ follows where $|A| = ad - bc$ (the determinant of A):

$$A^{-1} = \begin{bmatrix} d/|A| & -b/|A| \\ -c/|A| & a/|A| \end{bmatrix} = \frac{1}{|A|} \begin{bmatrix} d & -b \\ -c & a \end{bmatrix}.$$

That is, first find $|A|$. If $|A| = 0$, then A has no inverse. If $|A| \neq 0$, then A^{-1} is obtained by (i) interchanging a and d on the main diagonal, (ii) taking the negatives of the other elements b and c, and (iii) multiplying the matrix by $1/|A|$.

111. Find the inverse of each matrix:

(A) $A = \begin{bmatrix} 3 & 5 \\ 2 & 3 \end{bmatrix}$

(B) $A = \begin{bmatrix} 5 & 3 \\ 4 & 2 \end{bmatrix}$

(C) $A = \begin{bmatrix} -2 & 6 \\ 3 & -9 \end{bmatrix}$

Algorithm 4.1 (Gaussian Elimination Algorithm): The following algorithm finds the inverse of an n-square matrix A or determines that A has no inverse.

Step 1. Form the $n \times 2n$ (block) matrix $M = (A : I)$; that is, A is in the left half of M and I is in the right half of M.

Step 2. Row reduce M to echelon form. If the process generates a zero in the A-half of M, stop (A is not invertible). Otherwise, the A-half will assume triangular form.

Step 3. Further row reduce M to the row canonical form $(I : B)$, where I has replaced A in the left half of the matrix.

Step 4. Set $A^{-1} = B$.

112. Find the inverse of each matrix:

(A) $A = \begin{bmatrix} 1 & 0 & 2 \\ 2 & -1 & 3 \\ 4 & 1 & 8 \end{bmatrix}$

(B) $B = \begin{bmatrix} 1 & -2 & 2 \\ 2 & -3 & 6 \\ 1 & 1 & 7 \end{bmatrix}$

113. Find the inverse of $B = \begin{bmatrix} 1 & 3 & -4 \\ 1 & 5 & -1 \\ 3 & 13 & -6 \end{bmatrix}$.

114. Suppose A and B are invertible. Show that $A + B$ need not be invertible.

Elementary Matrices

Definition: Let e be an elementary row operation. The matrix $E = e(I)$ is called the *elementary matrix* corresponding to the elementary row operation e.

115. Find each 3-square elementary matrix:
    (A) E_1 corresponding to $R_1 \leftrightarrow R_3$
    (B) E_2 corresponding to $R_3 \rightarrow -7R_3$
    (C) E_3 corresponding to $R_2 \rightarrow -7R_1 + R_2$

Theorem 4.2: Let e be an elementary row operation and let E be the corresponding m-square elementary matrix, i.e., $E = e(I_m)$. Then for any $m \times n$ matrix A, $e(A) = EA$. That is, $e(A)$ may be obtained by pre-multiplying A by the corresponding elementary matrix E.

116. Show that the elementary matrices are invertible and that their inverses are also elementary matrices.

Theorem 4.3: The following are equivalent:
  (i) A is invertible.
 (ii) A is row equivalent to the identity matrix I.
(iii) A is a product of elementary matrices.

117. Suppose A is invertible and A is row reducible to the identity matrix I by a sequence of elementary row operations $e_1, e_2, \ldots, e_n$. Show that this sequence of elementary row operations applied to I yields A^{-1}.

118. Show that if AB is invertible, then A is invertible. (Thus, if A has no inverse, AB has no inverse.)

Special Matrices

Definition: A square matrix $A = [a_{ij}]$ is *upper triangular* if all entries below the diagonal are 0. Generic upper triangular matrices of order 2, 3, and 4 follow:

$$\begin{bmatrix} a_{11} & a_{12} \\ 0 & a_{22} \end{bmatrix}, \begin{bmatrix} b_{11} & b_{12} & b_{13} \\ 0 & b_{22} & b_{23} \\ 0 & 0 & b_{33} \end{bmatrix}, \begin{bmatrix} c_{11} & c_{12} & c_{13} & c_{14} \\ 0 & c_{22} & c_{23} & c_{24} \\ 0 & 0 & c_{33} & c_{34} \\ 0 & 0 & 0 & c_{44} \end{bmatrix}.$$

119. Suppose $A = [a_{ij}]$ and $B = [b_{ij}]$ are upper triangular. Show that

(A) $A + B$ is upper triangular with diagonal entries
$[a_{11} + b_{11}, a_{22} + b_{22}, \ldots, a_{nn} + b_{nn}]$
(B) kA is upper triangular with diagonal entries $[ka_{11}, ka_{22}, \ldots, ka_{nn}]$

120. Suppose $A = [a_{ij}]$ and $B = [b_{ij}]$ are upper triangular. Show that

(A) AB is upper triangular
(B) diagonal elements of AB are $a_{11}b_{11}, a_{22}b_{22}, \ldots, a_{nn}b_{nn}$

121. Using only 0 and 1, find all 2×2

(A) diagonal matrices,
(B) upper triangular matrices.

Definition: Let $A = [a_{ij}]$ be a real matrix.
(i) A is *symmetric* if $A^T = A$ or if each $a_{ij} = a_{ji}$.
(ii) A is *skew-symmetric* or *anti-symmetric* if $A^T = -A$ or if each $a_{ij} = -a_{ji}$.

122. Show that the diagonal elements of a skew-symmetric matrix must be 0.

123. Determine whether each matrix is symmetric or skew-symmetric:

(A) $A = \begin{bmatrix} 0 & 5 & -2 \\ -5 & 0 & 3 \\ 2 & -3 & 0 \end{bmatrix}$

(B) $B = \begin{bmatrix} 4 & -7 & 1 \\ -7 & 3 & 2 \\ 1 & 2 & -5 \end{bmatrix}$

(C) $C = \begin{bmatrix} 1 & 1 & 1 \\ 1 & 1 & 1 \end{bmatrix}$

124. Find x and A if $A = \begin{bmatrix} 4 & x+2 \\ 2x-3 & x+1 \end{bmatrix}$ is symmetric.

Theorem 4.4: Let A be a square matrix. Then

(i) $A + A^T$ is symmetric.

(ii) $A - A^T$ is skew-symmetric.

(iii) $A = B + C$ for some symmetric matrix B and some skew-symmetric matrix C.

125. Write $A = \begin{bmatrix} 2 & 3 \\ 7 & 8 \end{bmatrix} = B + C$ where A is symmetric and C is skew-symmetric.

Orthogonal Matrices

Definition: A real matrix A is *orthogonal* if $AA^T = A^T A = I$. Thus A is square and $A^{-1} = A^T$.

126. Define an orthonormal set of vectors in $\mathbf{R}^n$.

127. Show that $A = \begin{bmatrix} a_1 & a_2 & a_3 \\ b_1 & b_2 & b_3 \\ c_1 & c_2 & c_3 \end{bmatrix}$ is orthogonal if and only if its rows

$\mathbf{u}_1 = (a_1, a_2, a_3)$, $\mathbf{u}_2 = (b_1, b_2, b_3)$, $\mathbf{u}_3 = (c_1, c_2, c_3)$ form an orthonormal set.

Theorem 4.5: Let A be a real matrix. Then the following are equivalent:

(i) A is orthogonal.

(ii) The rows of A form an orthonormal set.

(iii) The columns of A form an orthonormal set.

128. Find x and y if $A = \begin{bmatrix} 1/\sqrt{5} & 2/\sqrt{5} \\ x & y \end{bmatrix}$ is orthogonal.

129. Show that the most general 2×2 orthogonal matrix has the form $\begin{bmatrix} a & b \\ b & -a \end{bmatrix}$ or $\begin{bmatrix} a & b \\ -b & a \end{bmatrix}$.

130. Show that every 2×2 orthogonal matrix has the form $\begin{bmatrix} \cos\theta & \sin\theta \\ -\sin\theta & \cos\theta \end{bmatrix}$ or $\begin{bmatrix} \cos\theta & \sin\theta \\ \sin\theta & -\cos\theta \end{bmatrix}$ for some real number θ.

Square Block Matrices

Definition: A block matrix A is a *square block matrix* if
 (i) A is a square matrix,
 (ii) the blocks form a square matrix, and
(iii) the diagonal blocks are square matrices.

131. Consider the following block matrices A, B, C:

$$A = \begin{bmatrix} 1 & 2 & 3 & 4 & 5 \\ 1 & 1 & 1 & 1 & 1 \\ 9 & 8 & 7 & 6 & 5 \\ 3 & 3 & 3 & 3 & 3 \\ 1 & 3 & 5 & 7 & 9 \end{bmatrix}, \quad B = \begin{bmatrix} 1 & 2 & 3 & 4 & 5 \\ 1 & 1 & 1 & 1 & 1 \\ 9 & 8 & 7 & 6 & 5 \\ 3 & 3 & 3 & 3 & 3 \\ 1 & 3 & 5 & 7 & 9 \end{bmatrix}, \quad C = \begin{bmatrix} 1 & 2 & 3 & 4 & 5 \\ 1 & 1 & 1 & 1 & 1 \\ 9 & 8 & 7 & 6 & 5 \\ 3 & 3 & 3 & 3 & 3 \\ 1 & 3 & 5 & 7 & 9 \end{bmatrix}.$$

(A) Is A a square block matrix?
(B) Is B a square block matrix?
(C) Complete the partitioning of C into a square block matrix.

Questions 132–134 refer to the following matrices:

$$A = \begin{bmatrix} 1 & 0 & 0 \\ 0 & 0 & 2 \\ 0 & 0 & 3 \end{bmatrix}, \quad B = \begin{bmatrix} 1 & 2 & 0 & 0 & 0 \\ 3 & 0 & 0 & 0 & 0 \\ 0 & 0 & 4 & 0 & 0 \\ 0 & 0 & 5 & 0 & 0 \\ 0 & 0 & 0 & 0 & 6 \end{bmatrix}, \quad C = \begin{bmatrix} 0 & 1 & 0 \\ 0 & 0 & 0 \\ 0 & 2 & 0 \end{bmatrix}.$$

132. Partition A into a block diagonal matrix with as many diagonal blocks as possible.

133. Partition B into a block diagonal matrix with as many diagonal blocks as possible.

134. Partition C into a block diagonal matrix with as many diagonal blocks as possible.

135. Let $M = \text{diag}(A_1, A_2, \ldots, A_r)$ and $N = \text{diag}[B_1, B_2, \ldots, B_r]$ be diagonal block matrices where corresponding diagonal blocks have the same size. Find:

(A) $M + N$
(B) kM
(C) MN
(D) $f(M)$ for a given polynomial $f(x)$

CHAPTER 5

Determinants

Definition: Each *n*-square matrix $A = [a_{ij}]$ is assigned a special scalar called the *determinant* of A, denoted by $\det(A)$ or $|A|$. The *order* of the determinant is the integer n.

Determinants of Order One and Two

136. The determinant function is said to be multiplicative. What does this mean?

137. The determinant of a 1×1 matrix $A = [a_{11}]$ is the scalar a_{11} itself. Find $\det(24)$, $\det(-6)$, $\det(t + 2)$.

Definition: The determinant of a 2×2 matrix $A = [a_{ij}]$ follows:

$$\det(A) = \begin{vmatrix} a_{11} & a_{12} \\ a_{21} & a_{22} \end{vmatrix} = \begin{vmatrix} a_{11} & a_{12} \\ a_{21} & a_{22} \end{vmatrix} = a_{11}a_{22} - a_{12}a_{21}.$$

That is, $|A|$ equals the product of the elements along the plus-labeled arrow of the diagram minus the product of the elements along the minus-labeled arrow.

138. Find the determinant of each matrix:

(A) $A = \begin{bmatrix} 5 & 4 \\ 2 & 3 \end{bmatrix}$
(B) $B = \begin{bmatrix} 2 & 1 \\ -4 & 6 \end{bmatrix}$

(C) $C = \begin{bmatrix} 3 & -2 \\ 4 & 5 \end{bmatrix}$
(D) $D = \begin{bmatrix} 4 & -5 \\ -1 & -2 \end{bmatrix}$

(E) $E = \begin{bmatrix} a & b \\ c & d \end{bmatrix}$

139. Determine those values of t for which $\begin{bmatrix} t-2 & 3 \\ 4 & t-1 \end{bmatrix} = 0$.

Determinants of Order Three

Definition: The determinant of a 3-square matrix $A = [a_{ij}]$ follows:

$$\det(A) = \begin{vmatrix} a_{11} & a_{12} & a_{13} \\ a_{21} & a_{22} & a_{23} \\ a_{31} & a_{32} & a_{33} \end{vmatrix} = a_{11}a_{22}a_{33} + a_{12}a_{23}a_{31} + a_{13}a_{21}a_{32} - a_{13}a_{22}a_{31}$$
$$- a_{12}a_{21}a_{33} - a_{11}a_{23}a_{32}.$$

Remark 5.1: Observe that there are six products, each product consisting of three elements of the original matrix coming from different rows and from different columns. Three products are plus-labeled (keep their sign) and three products are minus-labeled (change their sign).

Remark 5.2: The diagrams in Figure 5.1 may help you remember the six products in det(A). That is, det(A) is equal to the sum of the products along the three plus-labeled arrows in Figure 5.1 plus the sum of the negatives of the products along the three minus-labeled arrows. (We emphasize that there are no such diagrammatic devices with which to remember determinants of higher order.)

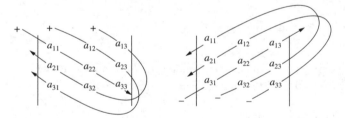

Figure 5.1

140. Use Figure 5.1 to find the determinant of each matrix:

(A) $A = \begin{bmatrix} 2 & 1 & 1 \\ 0 & 5 & -2 \\ 1 & -3 & 4 \end{bmatrix}$.

(B) $B = \begin{bmatrix} 3 & -2 & -4 \\ 2 & 5 & -1 \\ 0 & 6 & 1 \end{bmatrix}$.

(C) $C = \begin{bmatrix} -2 & -1 & 4 \\ 6 & -3 & -2 \\ 4 & 1 & 2 \end{bmatrix}$.

Alternative Form for a Determinant of Order Three

Remark 5.3: The determinant of a 3-square matrix $A = [a_{ij}]$ may be written as a linear combination of determinants of order two with coefficients from the first row as follows:

$$|A| = \begin{vmatrix} a_{11} & a_{12} & a_{13} \\ a_{21} & a_{22} & a_{23} \\ a_{31} & a_{32} & a_{33} \end{vmatrix} = a_{11} \begin{vmatrix} a_{11} & a_{12} & a_{13} \\ a_{21} & a_{22} & a_{23} \\ a_{31} & a_{32} & a_{33} \end{vmatrix} - a_{12} \begin{vmatrix} a_{11} & a_{12} & a_{13} \\ a_{21} & a_{22} & a_{23} \\ a_{31} & a_{32} & a_{33} \end{vmatrix}$$

$$+ a_{13} \begin{vmatrix} a_{11} & a_{12} & a_{13} \\ a_{21} & a_{22} & a_{23} \\ a_{31} & a_{32} & a_{33} \end{vmatrix} = a_{11} \begin{vmatrix} a_{22} & a_{23} \\ a_{32} & a_{33} \end{vmatrix} - a_{12} \begin{vmatrix} a_{21} & a_{23} \\ a_{31} & a_{33} \end{vmatrix} + a_{13} \begin{vmatrix} a_{21} & a_{22} \\ a_{31} & a_{32} \end{vmatrix}.$$

Note that each 2×2 matrix is obtained by deleting, in the original matrix, the row and column containing its coefficient (an element of the first row). Note also that the coefficients are taken with alternating signs.

141. Expand by the first row to find the determinant of each matrix.

(A) $A = \begin{bmatrix} 1 & 2 & 3 \\ 4 & -2 & 3 \\ 0 & 5 & -1 \end{bmatrix}$.

(B) $B = \begin{bmatrix} 2 & 0 & 1 \\ 3 & 2 & -3 \\ -1 & -3 & 5 \end{bmatrix}$.

(C) $C = \begin{bmatrix} 1 & 0 & 0 \\ 3 & 2 & -4 \\ 4 & 1 & 3 \end{bmatrix}$.

Remark 5.4: Let $A = \begin{bmatrix} a_1 & b_1 & c_1 \\ a_2 & b_2 & c_2 \\ a_3 & b_3 & c_3 \end{bmatrix}$. There are analogous expansions for $\det(A)$ using other rows as follows:

$$|A| = -a_2 \begin{vmatrix} b_1 & c_1 \\ b_3 & c_3 \end{vmatrix} + b_2 \begin{vmatrix} a_1 & c_1 \\ a_3 & c_3 \end{vmatrix} - c_2 \begin{vmatrix} a_1 & b_1 \\ a_3 & b_3 \end{vmatrix} \qquad \text{second row}$$

$$= a_3 \begin{vmatrix} b_1 & c_1 \\ b_2 & c_2 \end{vmatrix} - b_3 \begin{vmatrix} a_1 & c_1 \\ a_2 & c_2 \end{vmatrix} + c_3 \begin{vmatrix} a_1 & b_1 \\ a_2 & b_2 \end{vmatrix}. \qquad \text{third row}$$

Note that the signs of the coefficients form a checkerboard pattern as follows:

$$\begin{bmatrix} + & - & + \\ - & + & - \\ + & - & + \end{bmatrix}.$$

One can also expand the determinant so that the coefficients come from a column rather than a row, and the signs of the coefficients follow the same checkerboard pattern.

Determinants and Systems of Linear Equations

142. Solve the system using determinants: $3y + 2x = z + 1$, $3x + 2z = 8 - 5y$, $3z - 1 = x - 2y$.

Permutations

Definition: Let $S = \{1, 2, \ldots, n\}$. A *permutation* σ of S is a one-to-one mapping of S into S or, equivalently, a rearrangement of the numbers $1, 2, \ldots, n$. Such a permutation σ is denoted by

$$\sigma = \begin{pmatrix} 1 & 2 & \cdots & n \\ j_1 & j_2 & \cdots & j_n \end{pmatrix} \quad \text{or} \quad \sigma = j_1 j_2 \cdots j_n$$

where $ji = \sigma(i)$. There are $n!$ such permutations σ, denoted by S_n.

143. List the permutations in S_2.

144. List the permutations in S_3.

Sign (Parity) of a Permutation

Definition: Let σ be a permutation in S_n, say $\sigma = j_1 j_2 \cdots j_n$. We say the parity of σ is *even* or *odd* depending on whether there is an even or odd number of inversions in σ. By an *inversion* in σ we mean a pair of integers (i, k) such that $i > k$ but i precedes k in σ. The sign of σ, written $\text{sign}(\sigma)$, is defined by

$$\text{sign}(\sigma) = \begin{cases} 1 & \text{if } \sigma \text{ is even.} \\ -1 & \text{if } \sigma \text{ is odd.} \end{cases}$$

145. Find the sign of $\sigma = 35142$ in S_5.

146. Find the sign of the identity permutation $\varepsilon = 123 \cdots n$.

147. Find the signs of the permutations in

  (A) S_2
  (B) S_3

Remark 5.5: We note that half of the permutations in S_n are even and half are odd.

Determinants of Arbitrary Order

Remark 5.6: Let $A = [a_{ij}]$ be an n-square matrix. Consider a product of n elements of A such that no two elements are in the same row or the same column. Such a product can be written in the form

$$a_{1j_1} a_{2j_2} \cdots a_{nj_n},$$

where we have chosen to order the factors by rows, making the sequence of second subscripts a permutation $\sigma = j_1 j_2 \ldots j_n$ in S_n.

Definition: The determinant of $A = [a_{ij}]$, denoted by $\det(A)$ or $|A|$, is the sum of all the above $n!$ products, where each product is multiplied by sign σ. That is,

$$|A| = \sum_\sigma (\text{sign } \sigma) a_{1j_1} a_{2j_2} \cdots a_{nj_n} = \sum_{\sigma \in S_n} (\text{sign } \sigma) a_{1\sigma(1)} a_{2\sigma(2)} \cdots a_{n\sigma(n)}.$$

Such a determinant is said to be of order n.
By Remark 5.5, half the products carry a plus sign (+), and half carry a minus sign (−).

Theorem 5.1: Let B be obtained from a square matrix A by an elementary row operation.

  (i)   If two rows of A were interchanged, then $|B| = -|A|$.
  (ii) If a row of A was multiplied by a scalar k, then $|B| = k\,|A|$.
  (iii) If a multiple of a row was added to another row, then $|B| = |A|$.

Theorem 5.2: For a square matrix A, the following are equivalent:

  (i) A is invertible.
  (ii) A is nonsingular.
  (iii) $AX = 0$ has only the zero solution.

Theorem 5.3: Let A and B be n-square matrices. Then $\det(A) = \det(A)\det(B)$.

148. Suppose P is nonsingular. Show that $|P^{-1}| = |P|^{-1}$.

149. Suppose B is similar to A; that is, $B = P^{-1}AP$. Show that $|B| = |A|$.

Evaluation of Determinants, Laplace Expansion

Definition: Let $A = [a_{ij}]$ be an n-square matrix. Let M_{ij} be the $(n-1)$-square submatrix of A obtained by deleting the ith row and the jth column of A. Then the determinant $|M_{ij}|$ is called the ij-minor of A, and the signed minor $A_{ij} = (-1)^{i+j}|M_{ij}|$ is called the ij-cofactor of A.

150. Let $A = \begin{bmatrix} 2 & 3 & 4 \\ 5 & 6 & 7 \\ 8 & 9 & 1 \end{bmatrix}$.

    (A) Find M_{21} and the minor $|M_{21}|$ of A.
    (B) Find the cofactor A_{21}.
    (C) Find the minor $|M_{22}|$ of A.
    (D) Find the cofactor A_{22}.
    (E) Find the minor $|M_{23}|$ of A.
    (F) Find the cofactor A_{23}.

Theorem 5.4 (Laplace Expansion Theorem): The *determinant* of the n-square matrix $A = [a_{ij}]$ is equal to the sum of the products obtained by multiplying the elements of any row (column) by their respective cofactors:

$$|A| = \sum_{j=1}^{n} a_{ij} A_{ij} \quad \text{and} \quad |A| = \sum_{i=1}^{n} a_{ij} A_{ij}.$$

Algorithm 5.1: The following algorithm reduces the determinant of an n-square matrix A to a determinant of order $n-1$.

Step 1. Choose an element $a_{ij} = 1$ or, lacking that, $a_{ij} \neq 0$.

Step 2. Using a_{ij} as a pivot, apply elementary row (column) operations to put 0s in all other positions in column j (row i).

Step 3. Expand the determinant using column j (row i).

If Step 2 involves multiplication of a row (column) by a scalar, the final answer must be adjusted according to Theorem 5.1(ii).

151. Use Algorithm 5.1 to obtain the determinant of each matrix.

(A) $A = \begin{bmatrix} 2 & 5 & -3 & -2 \\ -2 & -3 & 2 & -5 \\ 1 & 3 & -2 & 2 \\ -1 & -6 & 4 & 3 \end{bmatrix}$

(B) $B = \begin{bmatrix} 3 & -2 & -5 & 4 \\ 1 & -2 & -2 & 3 \\ -2 & 4 & 7 & -3 \\ 2 & -3 & -5 & 8 \end{bmatrix}$

Classical Adjoint

Definition: The classical *adjoint* of a square matrix A, denoted by adj(A), is the transpose of the matrix of cofactors of A.

152. Find adj(A) for the matrix $A = \begin{bmatrix} 2 & 3 & 4 \\ 5 & 6 & 7 \\ 8 & 9 & 1 \end{bmatrix}$.

Theorem 5.5: For any square matrix A,
 (i) $A \cdot \text{adj}(A) = \text{adj}(A) \cdot A = \text{diag}(|A|, |A|, \ldots, |A|) = |A|I$.
(ii) If $|A| \neq 0$, then $A^{-1} = (1/|A|) \cdot \text{adj}(A)$.

153. Let $A = \begin{bmatrix} 2 & 3 & -4 \\ 0 & -4 & 2 \\ 1 & -1 & 5 \end{bmatrix}$.

   (A) Find det(A).
   (B) Find adj(A).
   (C) Verify that $A \cdot \text{adj}(A) = |A|I$.
   (D) Use adj(A) to find A^{-1}.

Cramer's Rule, Block Matrices

Theorem 5.6 (Cramer's rule): Let $AX = B$ be an $n \times n$ system of linear equations with nonsingular coefficient matrix $A = (a_{ij})$. Let A_i be the matrix obtained from A by replacing the ith column of A by the column vector B. Let $D \equiv |A|$, and let $N_i = |A_i|$ for $i = 1, 2, \ldots, n$. Then the system has the unique solution $x_i = N_i/D$ ($i = 1, 2, \ldots, n$).

Remark 5.7: One rarely, if ever, uses Cramer's rule to solve a system of linear equations. It is much more efficient to use Gaussian elimination (Algorithm 3.2).

154. Use Cramer's rule to solve the system:

$$
\begin{aligned}
x_1 + x_2 + x_3 + x_4 &= 2 \\
x_1 + 2x_2 + 3x_3 + 4x_4 &= 2 \\
2x_1 + 3x_2 + 5x_3 + 9x_4 &= 2 \\
x_1 + x_2 + 2x_3 + 7x_4 &= 2.
\end{aligned}
$$

Remark 5.8: Suppose $M = \begin{bmatrix} A & C \\ 0 & B \end{bmatrix}$ is an upper triangular square block matrix. Then $\det(M) = \det(A)\det(B)$.

155. Find $|M|$ where $M = \begin{bmatrix} 2 & 3 & 4 & 7 & 8 \\ -1 & 5 & 3 & 2 & 1 \\ 0 & 0 & 2 & 1 & 5 \\ 0 & 0 & 3 & -1 & 4 \\ 0 & 0 & 5 & 2 & 6 \end{bmatrix}$.

Minors

156. Let $A = [a_{ij}]$ be a 5-square matrix. Find the minor $|A_{3,5}^{1,4}|$ and its signed minor.

157. When is a minor a *principal minor*?

158. Let $A = [a_{ij}]$ be a 5-square matrix. Consider the following minors of A:

$$M_1 = \begin{vmatrix} a_{22} & a_{24} & a_{25} \\ a_{42} & a_{44} & a_{45} \\ a_{52} & a_{54} & a_{55} \end{vmatrix}, \quad M_2 = \begin{vmatrix} a_{11} & a_{13} & a_{15} \\ a_{21} & a_{23} & a_{25} \\ a_{51} & a_{53} & a_{55} \end{vmatrix}, \quad M_3 = \begin{vmatrix} a_{22} & a_{25} \\ a_{52} & a_{55} \end{vmatrix}.$$

Decide whether each minor is a principal minor.

    (A) M_1
    (B) M_2
    (C) M_3

159. Find the complement of the above minor M_1.

160. Find the complement of the above minor M_2.

CHAPTER 6

Vector Spaces

Here: (i) V is a vector space over K; (ii) **u**, **v**, **w** are elements of V; (iii) a, b, c, k are scalars in K.

Vector Spaces

Definition: Let K be a given field and let V be a nonempty set with rules of addition and scalar multiplication that assign to any **u**, **v** $\in$ V a *sum* **u** + **v** $\in$ V and to any **u** $\in$ V, $k \in K$ a *product* k**u** $\in$ V. Then V is called a *vector space over K* (and the elements of V are called *vectors*) if the following axioms hold.

[A_1]: For any vectors **u**, **v**, **w** $\in$ V, (**u** + **v**) + **w** = **u** + (**v** + **w**).

[A_2]: There is a vector in V, denoted by **0** and called the *zero vector*, for which **u** + **0** = **u** for any vector **u** $\in$ V.

[A_3]: For each vector **u** $\in$ V there is a vector in V, denoted by $-$**u**, for which **u** + ($-$**u**) = **0**.

[A_4]: For any vectors **u**, **v** $\in$ V, **u** + **v** = **v** + **u**.

[M_1]: For any scalar $k \in K$ and any vectors **u**, **v** $\in$ V, k(**u** + **v**) = k**u** + k**v**.

[M_2]: For any scalars a, $b \in K$ and any vector **u** $\in$ V, $(a + b)$**u** = a**u** + b**u**.

[M_3]: For any scalars a, $b \in K$ and any vector **u** $\in$ V, (ab)**u** = $a(b$**u**$)$.

[M_4]: For the unit scalar $1 \in K$, 1**u** = **u** for any vector **u** $\in$ V.

161. In the statement of axiom [M_2], $(a + b)$**u** = a**u** + b**u**, which operation does each plus sign represent?

162. Let $V = K^n$. Show how V is made into a vector space over K.

163. Let V be the set of $m \times n$ matrices in K. Show how V is made into a vector space over K.

164. Let V be the set of polynomials in K. Show how V is made into a vector space over K.

Theorem 6.1: Let V be a vector space over a field K.
 (i) For any scalar $k \in K$ and $\mathbf{0} \in V$, $k\mathbf{0} = \mathbf{0}$.
 (ii) For $0 \in K$ and any vector $\mathbf{u} \in V$, $0\mathbf{u} = \mathbf{0}$.
 (iii) If $k\mathbf{u} = \mathbf{0}$, where $k \in K$ and $\mathbf{u} \in V$, then $k = 0$ or $\mathbf{u} = \mathbf{0}$.
 (iv) For any $k \in K$ and any $\mathbf{u} \in V$, $(-k)\mathbf{u} = k(-\mathbf{u}) = -k\mathbf{u}$.

165. Prove Theorem 6.1:
  (A) $k\mathbf{0} = \mathbf{0}$
  (B) $0\mathbf{u} = \mathbf{0}$
  (C) If $k\mathbf{u} = \mathbf{0}$, then $k = 0$ or $\mathbf{u} = 0$
  (D) $(-k)\mathbf{u} = k(-\mathbf{u}) = -k\mathbf{u}$

Subspaces

Definition:: A subset W of a vector space V is called a *subspace* of V if W itself is a vector space with respect to the operations of vector addition and scalar multiplication in V.

Theorem 6.2: W is a subspace of V if (i) $0 \in W$ (or $W \neq \varnothing$) and (ii) if $\mathbf{v}, \mathbf{w} \in W$, then $a\mathbf{v} + b\mathbf{w} \in W$.

166. Show that the subset W of vectors whose sum of components equals zero is a subspace of $\mathbf{R}^3$.

167. Show that the subset W of symmetric matrices is a subspace the set V of n-square matrices.

Theorem 6.3: The solution set W of a homogeneous system in n unknowns is a subspace of K^n.

168. Prove Theorem 6.3.

Linear Combinations, Linear Spans, Row Space

Definition: Any vector in V of the form $a_1\mathbf{v}_1 + a_2\mathbf{v}_2 + \cdots + a_m\mathbf{v}_m$ is called a *linear combination* of the $\mathbf{v}$'s. The *linear span* of a subset S of V, denoted by span(S) or $L(S)$, consists of all linear combinations of elements in S. If $S = \varnothing$, then $L(S) = \{0\}$.

169. Write $\mathbf{v} = (1, -2, 5)$ as a linear combination of $\mathbf{e}_1 = (1, 1, 1)$, $\mathbf{e}_2 = (1, 2, 3)$, $\mathbf{e}_3 = (2, -1, 1)$.

170. For which value of k will $\mathbf{u} = (1, -2, k)$ be a linear combination of $\mathbf{v} = (3, 0, -2)$ and $\mathbf{w} = (2, -1, -5)$?

171. Show that $\mathbf{e}_1 = (1, 0, 0)$, $\mathbf{e}_2 = (0, 1, 0)$, $\mathbf{e}_3 = (0, 0, 1)$ span $\mathbf{R}^3$.

Definition: Let A be an $m \times n$ matrix with rows $R_1, \ldots, R_m$ and columns $C_1, \ldots, C_n$. Then the *row space* of A is denoted and defined by $\mathrm{rowsp}(A) = \mathrm{span}(R_1, \ldots, R_m)$, and the column space of A is denoted and defined by $\mathrm{colsp}(A) = \mathrm{span}(C_1, \ldots, C_n)$.

Theorem 6.4:

(i) Row equivalent matrices have the same row space.

(ii) Any matrix A is row equivalent to a unique matrix in row canonical form.

(iii) Matrices A and B have the same row space if and only if their row canonical forms have the same nonzero rows.

172. Determine whether the following matrices have the same row space:

$$A = \begin{bmatrix} 1 & 1 & 5 \\ 2 & 3 & 13 \end{bmatrix}, \quad B = \begin{bmatrix} 1 & -1 & -2 \\ 3 & -2 & -3 \end{bmatrix}, \quad C = \begin{bmatrix} 1 & -1 & -1 \\ 4 & -3 & -1 \\ 3 & -1 & 3 \end{bmatrix}.$$

173. Let $\mathbf{r}$ be a row vector and B be a matrix such that $\mathbf{r}B$ is defined. Show that $\mathbf{r}B$ is a linear combination of the rows of B.

174. Let U be the span of $\mathbf{u}_1 = (1, 1, -1)$, $\mathbf{u}_2 = (2, 3, -1)$, $\mathbf{u}_3 = (3, 1, -5)$, and let W be the span of $\mathbf{w}_1 = (1, -1, -3)$, $\mathbf{w}_2 = (3, -2, -8)$, $\mathbf{w}_3 = (2, 1, -3)$. Show that $U = W$.

Sums and Direct Sums

Remark 6.1: $U + W$ consists of all sums $\mathbf{u} + \mathbf{w}$ where $\mathbf{u} \in U$ and $\mathbf{w} \in W$.

175. Suppose U and W are subspaces of V. Show that $U + W$ is a subspace of V.

Definition: V is the *direct sum* of its subspaces U and W, written $V = U \oplus W$, if every vector $\mathbf{v} \in V$ can be written in one and only one way as $\mathbf{v} = \mathbf{u} + \mathbf{w}$ where $\mathbf{u} \in U$ and $\mathbf{w} \in W$.

Theorem 6.5: V is the direct sum of subspaces U and W if

(i) $V = U + W$ and

(ii) $U \cap W = \{0\}$.

176. In $\mathbf{R}^3$, let U be the xy-plane and let W be the z-axis. Show $\mathbf{R}^3 = U \oplus W$.

177. Consider subspaces $U = \{(a, b, c) \mid a = b = c\}$ and $W = \{(0, b, c)\}$ of $\mathbf{R}^3$. Show that $U \oplus W$.

Linear Dependence and Independence

Definition: The vectors $v_1, \ldots, v_m \in V$ are said to be *linearly dependent over K*, or simply *dependent*, if there exist scalars $a_1, \ldots, a_m \in K$, not all of them 0, such that

$$a_1 v_1 + a_2 v_2 + \cdots + a_m v_m = 0.$$

Otherwise, the vectors are said to be *linearly independent over K*, or simply *independent*.

178. Show that two vectors are dependent if and only if one is a multiple of the other.

179. Describe geometrically the linear dependence of two vectors and three vectors in $\mathbf{R}^3$.

180. Determine whether **u** and **v** are linearly dependent where:
   (A) $u = (4, -3, 2)$, $v = (2, -6, 7)$
   (B) $u = (-4, 6, -2)$, $v = (2, -3, 1)$

181. Determine whether $(1, -2, 1)$, $(2, 1, -1)$, $(7, -4, 1)$ are linearly dependent.

182. Determine whether $(1, 2, -3)$, $(1, -3, 2)$, $(2, -1, 5)$ are linearly dependent.

183. Suppose **u**, **v**, **w** are independent. Show that $u + v$, $u - v$, $u - 2v + w$ are independent.

Lemma 6.6: The nonzero vectors $v_1, v_2, \ldots, v_m$ are linearly dependent if and only if one of them is a linear combination of the preceding vectors.

Theorem 6.7: The nonzero rows $R_1, R_2, \ldots, R_n$ of a matrix in echelon form are independent.

Basis and Dimension

Definition: A list $S = \{u_1, u_2, \ldots, u_n\}$ is a *basis* of V if (i) S is independent and (ii) S spans V.

Theorem 6.8: Every basis of V has the same number of elements.

Definition: If V has a basis with n elements, then n is called the *dimension* of V, written $\dim(V) = n$. If $V = \{0\}$, then $\dim(V) = 0$.

184. Consider the following n vectors in $\mathbf{R}^n$:

$$\mathbf{e}_1 = (1, 0, 0, \ldots, 0, 0), \mathbf{e}_2 = (0, 1, 0, \ldots, 0, 0), \ldots, \mathbf{e}_n = (0, 0, \ldots, 0, 1).$$

(A) Show that the vectors form a basis of $\mathbf{R}^n$.
(B) Show that $\dim(\mathbf{R}^n) = n$.

185. Determine whether $(1, 1, 1)$, $(1, 2, 3)$, $(2, -1, 1)$ form a basis of $\mathbf{R}^3$.

186. Determine whether $(1, 1, 2)$, $(1, 2, 5)$, $(5, 3, 4)$ form a basis of $\mathbf{R}^3$.

187. Show that $\mathbf{C}$ is a vector space of dimension two over $\mathbf{R}$.

188. Show that $\mathbf{R}$ is a vector space of infinite dimension over the rational field $\mathbf{Q}$.

Dimension and Subspaces

Theorem 6.9: Let W be a subspace of V. Then $\dim(W) \leq \dim(V)$. If $\dim(W) = \dim(V)$, then $W = V$.

189. Let W be a subspace of $\mathbf{R}^3$. Give a geometrical description of W in terms of its dimension.

190. Consider the subspace $W = \{(a, b, c) \mid a + b + c = 0\}$ of $\mathbf{R}^3$. Find a basis and dimension of W.

191. Let W be spanned by $(1, 4, -1, 3)$, $(2, 1, -3, -1)$, $(0, 2, 1, -5)$ in $\mathbf{R}^4$. Find a basis and dimension of W.

192. Let W be spanned by $(1, -4, -2, 1)$, $(1, -3, -1, 2)$, $(3, -8, -2, 7)$ in $\mathbf{R}^4$. Find a basis and dimension of W.

Rank of a Matrix

Definition: The *rank*, or *row rank*, of a matrix A, denoted by rank(A), is equal to the maximum number of independent rows of A or, equivalently, the dimension of the row space of A.

193. Find the rank of $A = \begin{bmatrix} 1 & 2 & 0 & -1 \\ 2 & 6 & -3 & -3 \\ 3 & 10 & -6 & -5 \end{bmatrix}$.

194. Find the rank of $B = \begin{bmatrix} 1 & 3 & 1 & -2 & -3 \\ 1 & 4 & 3 & -1 & -4 \\ 2 & 3 & -4 & -7 & -3 \\ 3 & 8 & 1 & -7 & -8 \end{bmatrix}$.

Theorem 6.10: The row rank and column rank of any matrix are equal.

195. Find the rank of $A = \begin{bmatrix} 1 & 2 & -3 \\ 2 & 1 & 0 \\ -2 & -1 & 3 \\ -1 & 4 & -2 \end{bmatrix}$.

196. Let A and B be arbitrary matrices for which the product AB is defined. Show that:

(A) $\operatorname{rank}(AB) \leq \operatorname{rank}(B)$

(B) $\operatorname{rank}(AB) \leq \operatorname{rank}(A)$

Applications to Linear Equations

Algorithm 6.1: Suppose a homogeneous system $AX = 0$ is in echelon form with n unknowns and r equations. The following algorithm finds a basis and dimension of the solution space W of $AX = 0$.

Step 1. Find the $n - r$ free variables $x_{i_1}, x_{i_2}, \ldots, x_{i_{n-r}}$.

Step 2. Find the solution $\mathbf{v}_j$ obtained by setting $x_{i_j} = 1$ (or any $k \neq 0$) and the remaining free variables equal to 0.

Step 3. The solutions $\mathbf{v}_1, \mathbf{v}_2, \ldots, \mathbf{v}_{n-r}$ form a basis of W, and so $\dim(W) = n - r$.

197. Find the dimension and a basis of the solution subspace W of the system

$$x + 2y + 2z - s + 3t = 0, \quad x + 2y + 3z + s + t = 0, \quad 3x + 6y + 8z + s + 5t = 0.$$

198. Find the dimension and a basis of the solution subspace W of the system

$$x + 2y + z - 3t = 0, \quad 2x + 4y + 4z - t = 0, \quad 3x + 6y + 7z + t = 0.$$

199. Find the dimension and a basis of the solution subspace W of the system

$$x + 2y - 3z = 0, \quad 2x + 5y + z = 0, \quad x - y + 2z = 0.$$

200. Find a homogeneous system whose solution space W is spanned by

$$\{(1, -2, 0, 3), (1, -1, -1, 4), (1, 0, -2, 5)\}.$$

Sums, Direct Sums, Intersections

Theorem 6.11:
(i) $\dim(U + W) = \dim(U) + \dim(W) - \dim(U \cap W)$.
(ii) $\dim(U \cap W) = \dim(U) + \dim(W) - \dim(U + W)$.

201. Suppose U and W are subspaces of V where $\dim(U) = 4$, $\dim(W) = 4$, and $\dim(V) = 6$. Find the possible dimensions of $U \cap W$.

202. Suppose U and W are subspaces of $\mathbf{R}^3$ where $\dim(U) = 2$ and $\dim(W) = 2$. Show that $U \cap W \neq \emptyset$, and find the possible dimensions of $U \cap W$.

203. Let $U = \text{span}\{(1, 1, 0, -1), (1, 2, 3, 0), (2, 3, 3, -1)\}$ and $W = \text{span}$ $\{(1, 2, 2, -2), (2, 3, 2, -3), (1, 3, 4, -3)\}$. Find a basis and dimension of:
(A) $U + W$
(B) U
(C) W
(D) Find $\dim(U \cap W)$

204. Suppose $V = U \oplus W$. Show that $\dim(V) = \dim(U) + \dim(W)$.

205. Let U and W be subspaces of $\mathbf{R}^3$ $\dim(U) = 1$, $\dim(W) = 2$, and $U \not\subseteq W$. Show that $\mathbf{R}^3 = U \oplus W$.

Coordinate Vectors

Definition: Let $e = \{\mathbf{e}_1, \mathbf{e}_2, \ldots, \mathbf{e}_n\}$ be a basis of V, and suppose $\mathbf{v} \in V$ such that

$$\mathbf{v} = a_1\mathbf{e}_1 + a_2\mathbf{e}_2 + \cdots + a_n\mathbf{e}_n.$$

The scalars $a_1, a_2, \ldots, a_n$ are called the *coordinates* of $\mathbf{v}$ in the basis $\mathbf{e}$. The *coordinate vector* of v in e is denoted and defined by $[\mathbf{v}]_e = [a_1, a_2, \ldots, a_n]$ (sometimes written as a column).

206. Let $\mathbf{v} = (3, 1, -4)$. Find the coordinate vector of v relative to each basis of $\mathbf{R}^3$:
(A) $\mathbf{f}_1 = (1, 1, 1)$, $\mathbf{f}_2 = (0, 1, 1)$, $\mathbf{f}_3 = (0, 0, 1)$
(B) $\mathbf{e}_1 = (1, 0, 0)$, $\mathbf{e}_2 = (0, 1, 0)$, $\mathbf{e}_3 = (0, 0, 1)$

207. For the basis $\mathbf{u}_1 = (2, 1)$, $\mathbf{u}_2 = (1, -1)$ of $\mathbf{R}^2$, find the coordinate vector of

(A) $\mathbf{v} = (2, 3)$
(B) $\mathbf{u} = (4, -1)$
(C) $\mathbf{w} = (3, -3)$
(D) $\mathbf{v} = (a, b)$

Change of Basis

Definition: Consider two bases $S = \{\mathbf{u}_i\}$ and $S'\{\mathbf{v}_i\}$ of V. Suppose, for $i = 1, \ldots, n$,

$$\mathbf{v}_i = a_{i1}\mathbf{u}_1 + a_{i2}\mathbf{u}_2 + \cdots + a_{in}\mathbf{u}_n.$$

Let P be the transpose of the above matrix of coefficients; that is, $P = [p_{ij}]$ where $p_{ij} = a_{ij}$. Then P is called the change-of-basis matrix from the "old" basis S to the "new" basis S'.

Theorem 6.12: Let P be the change-of-basis matrix from S to S'. Then $Q = P^{-1}$ is the change-of-basis matrix from S' to S.

Theorem 6.13: Let P be the change-of-basis matrix from S to S'. Then, for any $\mathbf{v}$ in V,

$$P[\mathbf{v}]_{S'} = [\mathbf{v}]_S \text{ and hence } P^{-1}[\mathbf{v}]_S = [\mathbf{v}]_{S'}.$$

Remark 6.2: That is, multiplying the coordinates of $\mathbf{v}$ in the "old" basis S by P^{-1} gives the coordinates of $\mathbf{v}$ in the "new" basis S'.

Questions 208–210 refer to bases $S = \{\mathbf{u}_1, \mathbf{u}_2\} = \{(1, 2), (3, 5)\}$ and $S' = \{\mathbf{v}_1, \mathbf{v}_2\} = \{(1, -1), (1, -2)\}$ of $\mathbf{R}^2$.

208. Find the change-of-basis matrix P from S to the "new" basis S'.

209. Find the change-of-basis matrix Q from the "new" basis S' back to the "old" basis S.

210. Let $\mathbf{v} = (2, 1)$.

(A) Find $[\mathbf{v}]_S$
(B) Find $[\mathbf{v}]_{S'}$
(C) Verify Theorem 6.13 that $P^{-1}[\mathbf{v}]_S = Q[\mathbf{v}]_S = [\mathbf{v}]_{S'}$

CHAPTER 7

Linear Mappings

Here we assume the reader is familiar with mappings (functions) $f: A \to B$ from the *domain set A* to the *target set B*, the *composition* mapping $g \circ f: A \to C$ where $f: A \to B$ and $g: B \to C$, and one-to-one, onto, and invertible mappings.

Linear Mappings

Definition: Let V and U be vector spaces over the same field K. A mapping $F: V \to U$ is called a *linear mapping* (or *linear transformation* or *vector space homomorphism*) if it satisfies the following two conditions:

 (i) For any $\mathbf{v}, \mathbf{w} \in V$, $F(\mathbf{v} + \mathbf{w}) = F(\mathbf{v}) + F(\mathbf{w})$.
(ii) For any $k \in K$ and any $\mathbf{v} \in V$, $F(k\mathbf{v}) = kF(\mathbf{v})$.

In other words, $F: V \to U$ is linear if it "preserves" the two basic operations of a vector space, that of vector addition and that of scalar multiplication.

211. Suppose $F: V \to U$ is linear. Show that:

    (A) $F(\mathbf{0}) = \mathbf{0}$
    (B) $F(-\mathbf{u}) = -F(\mathbf{u})$

212. Show that $F: V \to U$ is linear if and only if $F(a\mathbf{v} + b\mathbf{w}) = aF(\mathbf{v}) + bF(\mathbf{w})$. (This condition characterizes linear mapping and is sometimes used as its definition.)

213. Let A be an $m \times n$ matrix. Note that A determines a mapping $T: K^n \to K^m$ defined by $T(\mathbf{v}) = A\mathbf{v}$ (where $\mathbf{v}$ is a column vector). Show that T is linear.

214. Suppose $F: V \to U$ and $G: U \to W$ are linear. Show that the composition $G \circ F: V \to W$ is linear.

215. Suppose $F: V \to U$ is linear and one-to-one and onto. Show that the inverse $F^{-1}: U \to V$ is linear.

Theorem 7.1: Let V and U be vector spaces over a field K. Let $\{v_1, v_2, \ldots, v_n\}$ be a basis of V and let $u_1, u_2, \ldots, u_n$ be any vectors in U. Then there exists a unique linear mapping $F: V \rightarrow U$ such that $F(v_1) = u_1, F(v_2) = u_2, \ldots, F(v_n) = u_n$.

216. Consider the map $T: \mathbf{R}^2 \rightarrow \mathbf{R}^2$ for which $T(3, 1) = (2, -4)$ and $T(1, 1) = (0, 2)$.
  (A) Show that such a map is linear and is unique.
  (B) Find a formula for T.
  (C) Find $T(7, 4)$.
  (D) Find $T^{-1}(5, -3)$.

Kernel and Image of a Linear Mapping

Definition: Let $F: V \rightarrow U$ be linear. The *kernel* of F, written $\text{Ker}(F)$, is the set of vectors that map onto 0, and the *image* of F, written $\text{Im}(F)$, is the set of image points in U; that is;

$$\text{Ker}(F) = \{v \in V: F(v) = 0\} \text{ and } \text{Im}(F) = \{u \in U \mid \exists\, v \in V \text{ for which } F(v) = u\}$$

217. Suppose $F: V \rightarrow U$ is linear. Show that:
  (A) $\text{Ker}(F)$ is a subspace of V.
  (B) $\text{Im}(F)$ is a subspace of U.

218. Suppose $F: V \rightarrow U$ is linear and $v_1, v_2, \ldots, v_n$ span V. Show that $F(v_1), F(v_2), \ldots, F(v_n)$ span $\text{Im}(F)$.

Theorem 7.2: Let V be of finite dimension and let $F: V \rightarrow U$ be a linear mapping. Then

$$\dim(V) = \dim(\text{Ker}(F)) + \dim(\text{Im}(F)).$$

219. Let $T: \mathbf{R}^3 \rightarrow \mathbf{R}^3$ be defined by $T(x, y, z) = (x + 2y - z, y + z, x + y - 2z)$. Find a basis and dimension of each of the following:
  (A) the image U of T.
  (B) the kernel W of T.

220. Let $A: \mathbf{R}^4 \rightarrow \mathbf{R}^3$ be defined by $A = \begin{bmatrix} 1 & 2 & 3 & 1 \\ 1 & 3 & 5 & -2 \\ 3 & 8 & 13 & -3 \end{bmatrix}$.

  (A) Find a basis and dimension of the image of A.
  (B) Find the dimension of the kernel of A.
  (C) Find a basic of the kernel of A.

221. Let B: $\mathbf{R}^3 \to \mathbf{R}^3$ be defined by $B = \begin{bmatrix} 1 & 2 & 5 \\ 3 & 5 & 13 \\ -2 & -1 & -4 \end{bmatrix}$.

    (A) Find a basis and dimension of the kernel of B.
    (B) Find the dimension of the image of B.
    (C) Find a basis of the image of B.

Singular and Nonsingular Linear Mappings, Isomorphisms

Definition: A linear mapping F: $V \to U$ is *singular* if there exists $\mathbf{v} \neq \mathbf{0}$ in V for which $F(\mathbf{v}) = \mathbf{0}$. Thus F is *nonsingular* if $\mathrm{Ker}(F) = \{\mathbf{0}\}$.

222. Let F: $\mathbf{R}^2 \to \mathbf{R}^2$ be defined by $F(x, y) = (x - y, x - 2y)$. Is F nonsingular? If not, find $\mathbf{v} \neq \mathbf{0}$ such that $F(\mathbf{v}) = \mathbf{0}$.

223. Let G: $\mathbf{R}^2 \to \mathbf{R}^2$ be defined by $G(x, y) = (2x - 4y, 3x - 6y)$. Is G nonsingular? If not, find $\mathbf{v} \neq \mathbf{0}$ such that $G(\mathbf{v}) = \mathbf{0}$.

224. Let H: $\mathbf{R}^3 \to \mathbf{R}^3$ be defined by $H(x, y, z) = (x + y - 2z, x + 2y + z, 2x + 2y - 3z)$. Is H nonsingular? If not, find $\mathbf{v} \neq \mathbf{0}$ such that $H(\mathbf{v}) = \mathbf{0}$.

225. Let F: $\mathbf{R}^3 \to \mathbf{R}^3$ be defined by $F(x, y, z) = (x + y + z, x + 2y - z, 3x + 5y - z)$. Is F nonsingular? If not, find $\mathbf{v} \neq \mathbf{0}$ such that $F(\mathbf{v}) = \mathbf{0}$.

Definition: A linear map F: $V \to U$ is called an *isomorphism* if F is bijective—that is, one-to-one and onto. Vector space V is *isomorphic* to a vector space U, written $V \cong U$, if there is an isomorphism F: $V \to U$.

226. Suppose $\dim(V) = n$, Show that V is isomorphic to K^n.

227. The linear mapping H: $\mathbf{R}^3 \to \mathbf{R}^3$ defined by $H(x, y, z) = (x + y - 2z, x + 2y + z, 2x + 2y - 3z)$ is nonsingular (Problem 224). Find a formula for H^{-1}.

Operations with Linear Mappings

Definition: Let F: $V \to U$ and G: $V \to U$ be linear mappings. The *sum* $F + G$ and the *scalar product* kF are maps from V into U defined as follows:

$$(F + G)(\mathbf{v}) = F(\mathbf{v}) + G(\mathbf{v}) \text{ and } (kF)(\mathbf{v}) = kF(\mathbf{v}).$$

228. Suppose $F: V \to U$ and $G: V \to U$ are linear. Prove that:
(A) $F + G$ is linear
(B) kF is linear

229. Consider linear maps $F: \mathbf{R}^3 \to \mathbf{R}^2$ and $G: \mathbf{R}^3 \to \mathbf{R}^2$, defined by $F(x, y, z) = (2x, y + z)$ and $G(x, y, z) = (x - z, y)$, and $\mathbf{v} = (2, 3, 4)$. Find:
(A) $(F + G)(\mathbf{v})$
(B) $(3F)(\mathbf{v})$

230. Let F and G be the maps in Question 229. Find a formula for:
(A) $F + G$
(B) $3F$
(C) $2F - 5G$

231. Consider linear maps $F: \mathbf{R}^3 \to \mathbf{R}^2$, $G: \mathbf{R}^3 \to \mathbf{R}^2$, $H: \mathbf{R}^2 \to \mathbf{R}^2$ defined by

$$F(x, y, z) = (y, x + z), \quad G(x, y, z) = (2z, x - y), \quad H(x, y) = (y, 2x).$$

Find:
(A) $(H \circ F)(\mathbf{v})$ for $\mathbf{v} = (4, -1, 5)$
(B) $(H \circ G)(\mathbf{w})$ for $\mathbf{w} = (3, 4, 1)$

232. Let F, G, H be the linear maps in Question 231. Find a formula for:
(A) $F + G$
(B) $H \circ F$
(C) $H \circ G$
(D) $H \circ (F + G)$
(E) $H^2 = H \circ H$

Vector Space of Linear Mappings

Theorem 7.3: Let V and U be vector spaces over a field K. Then the collection of all linear mappings from V into U with the operations of addition and scalar multiplication of linear mappings form a vector space over K.

Remark 7.1: The notation $\text{Hom}(V, U)$ denotes the above vector space of linear mappings from V into U. (Here Hom comes from the word *homomorphism*.)

233. Consider linear maps $F: \mathbf{R}^3 \to \mathbf{R}^2$, $G: \mathbf{R}^3 \to \mathbf{R}^2$, $H: \mathbf{R}^3 \to \mathbf{R}^2$ defined by

$$F(x, y, z) = (x + y + z, x + y), \quad G(x, y, z) = (2x + z, x + y), \quad H(x, y, z) = (2y, x).$$

Find formulas for:

(A) $F + G$
(B) $F + H$
(C) $G \circ F$
(D) $3G + 2H$

234. Show that the linear maps F, G, H in Question 233 are linearly independent (as elements in Hom($\mathbf{R}^3$, $\mathbf{R}^2$)).

Theorem 7.4: Suppose dim(V) = m and dim(U) = n. Then dim(Hom(V, U)) = mn.

235. Find the dimension of Hom($\mathbf{R}^3$, $\mathbf{R}^2$).

Algebra of Linear Operators on V

Remark 7.2: Here we consider the special case of linear mappings $T: V \to V$, called *linear operators* or *linear transformations* on V. We write $A(V)$ instead of Hom(V, V) for the space of such mappings. The letter A is used since such mappings form an *algebra* over K.

236. Which of the integers 5, 9, 18, 25, 31, 36, 64, 88, 100 can be the dimension of an algebra $A(V)$?

237. Let T belong to $A(V)$.

(A) Define the powers T^2, T^3, $\cdots$ of T
(B) Define $p(T)$ for a polynomial $p(x) = a_0 + a_1 x + a_2 x^2 + \cdots + a_n x^n$ over K

238. Consider linear operators S, T in $A(\mathbf{R}^2)$ defined by $S(x, y) = (x + y, 0)$ and $T(x, y) = (-y, x)$. Find a formula for:

(A) $5S - 3T$
(B) ST
(C) TS

239. Consider linear operators S and T in Question 238. Show that:

(A) $S^2 = S$
(B) $T^2 = -I$

Questions 240–242 refer to the linear operator T in $A(\mathbf{R}^2)$ defined by $T(x, y) = (x + 2y, 3x + 4y)$.

240. Find a formula for
(A) T^2
(B) T^3

241. (A) Find $f(T)$ where $f(x) = x^2 - 3x + 4$.
(B) Is T a root of $f(x)$?

242. (A) Find $g(T)$ where $g(x) = x^2 - 5x - 2$.
(B) Is T a root of $g(x)$?

243. Let E be a linear operator in $A(V)$ for which $E^2 = E$ (called a projection). Show that $V = U + W$ where U is the image of E and W is the kernel of E.

Invertible Operators

Definition: A linear operator T in $A(V)$ is *invertible* if T has an *inverse*—that is, there exists T^{-1} in $A(V)$ such that $TT^{-1} = T^{-1}T = I$.

Theorem 7.5: A linear operator T on a finite-dimensional vector space V is invertible if and only if T is nonsingular.

244. Show that the finiteness of the dimensionality of V is necessary in Theorem 7.5.

245. Consider the linear operator T on $\mathbf{R}^2$ defined by $T(x, y) = (y, 2x - y)$.
(A) Show that T is invertible.
(B) Find a formula for T^{-1}.
(C) Find $T(6, 2)$.
(D) Find $T^{-1}(6, 2)$.

246. Consider the linear operator T on $\mathbf{R}^3$ defined by $T(x, y, z) = (2x, 4x - y, 2x + 3y - z)$.
(A) Show that T is invertible.
(B) Find a formula for T^{-1}.
(C) Find $T^{-1}(2, 4, 6)$.

247. Consider the linear operator T on $\mathbf{R}^3$ defined by $T(x, y, z) = (x + z, x - z, y)$.

    (A) Show that T is invertible.

    (B) Find a formula for T^{-1}.

    (C) Find $T^{-1}(2, 4, 6)$.

Remark 7.3: Two operators S and T in $A(V)$ are said to be *similar*, written $S \sim T$, if there exists an invertible operator P in $A(V)$ such that $S = P^{-1}TP$.

Questions 248–250 show that the relation $S \sim T$ is an equivalence.

248. [Reflexive] Show that $S \sim S$ for every S in $A(V)$.

249. [Symmetric] Suppose $S \sim T$. Show that $T \sim S$.

250. [Transitive] Suppose $F \sim G$ and $G \sim H$. Show that $F \sim H$.

CHAPTER 8

Matrices and Linear Mappings

Matrix Representation of a Linear Operator

Definition: Let $T: V \to V$ be linear and suppose $B = \{\mathbf{e}_1, \mathbf{e}_2, \ldots, \mathbf{e}_n\}$ is a basis of V. For $i = 1, \ldots, n$, suppose

$$T(\mathbf{v}_i) = a_{i1}\mathbf{e}_1 + a_{i2}\mathbf{e}_2 + \cdots + a_{in}\mathbf{e}_n.$$

The transpose of the above matrix of coefficients, denoted by $m_B(T)$ or $[T]_B$, is called the *matrix representation* of T relative to the basis B or simply the matrix T in the basis B.

Remark 8.1: All coordinate vectors are assumed to be column vectors unless otherwise stated or implied.

Theorem 8.1: Let $B = \{\mathbf{e}_1, \ldots, \mathbf{e}_n\}$ be a basis of V and T be any operator on V. Then, for any vector $\mathbf{v} \in V$, we have $[T]_B[\mathbf{v}]_B = [T(\mathbf{v})]_B$. [That is, if we multiply the coordinate vector of $\mathbf{v}$ by the matrix representation of T, then we obtain the coordinate vector of $T(\mathbf{v})$.]

Questions 251–253 refer to the map $F: \mathbf{R}^2 \to \mathbf{R}^2$ defined by $F(x, y) = (2x - 5y, 3x + y)$ and to the basis $B = \{\mathbf{u}_1 = (2, 1), \mathbf{u}_2 = (3, 2)\}$ of $\mathbf{R}^2$.

251. (A) Find $F(\mathbf{u}_1)$.
  (B) Write $F(\mathbf{u}_1)$ as a linear combination of $\mathbf{u}_1$ and $\mathbf{u}_2$.

252. (A) Find $F(\mathbf{u}_2)$.
  (B) Write $F(\mathbf{u}_2)$ as a linear combination of $\mathbf{u}_1$ and $\mathbf{u}_2$.

253. Find $[F]$, the matrix representation of F in the basis B.

Questions 254–257 refer to the map $G: \mathbf{R}^2 \to \mathbf{R}^2$ defined by $G(x, y) = (2x - 3y, 4x + y)$ and to the basis $B = \{\mathbf{u}_1 = (1, -2), \mathbf{u}_2 = (2, -5)\}$ of $\mathbf{R}^2$.

254. Find the coordinates of an arbitrary vector (a, b) with respect to the basis B.

255. (A) Find $G(\mathbf{u}_1)$.
(B) Write $G(\mathbf{u}_1)$ as a linear combination of $\mathbf{u}_1$ and $\mathbf{u}_2$.

256. (A) Find $G(\mathbf{u}_2)$.
(B) Write $G(\mathbf{u}_2)$ as a linear combination of $\mathbf{u}_1$ and $\mathbf{u}_2$.

257. Find $[G]_B$, the matrix representation of G in the basis B.

Questions 258–261 refer to $S: \mathbf{R}^2 \to \mathbf{R}^2$ defined by $S(x, y) = (2y, 3x - y)$ and to the basis $B = \{\mathbf{v}_1 = (1, 3), \mathbf{v}_2 = (2, 5)\}$ of $\mathbf{R}^2$.

258. Find the coordinates of an arbitrary vector (a, b) with respect to the basis B.

259. (A) Find $S(\mathbf{v}_1)$.
(B) Write $S(\mathbf{v}_1)$ as a linear combination of $\mathbf{v}_1$ and $\mathbf{v}_2$.

260. (A) Find $S(\mathbf{v}_2)$.
(B) Write $S(\mathbf{v}_2)$ as a linear combination of $\mathbf{v}_1$ and $\mathbf{v}_2$.

261. (A) Find $[S]_B$, the matrix representation of S in the basis B.
(B) Find $[S]_E$, the matrix representation of S in the usual basis
$E = \{\mathbf{e}_1 = (1, 0), \mathbf{e}_2 = (0, 1)\}$.

Problems 262–264 refer to $T: \mathbf{R}^2 \to \mathbf{R}^2$ defined by $T(x, y) = (3x - 4y, x + 5y)$ and to the above basis $B = \{\mathbf{v}_1 = (1, 3), \mathbf{v}_2 = (2, 5)\}$ of $\mathbf{R}^2$.
[**Note:** The coordinates of (a, b) in B are given in Problem 258.]

262. (A) Find $T(\mathbf{v}_1)$.
(B) Write $T(\mathbf{v}_1)$ as a linear combination of $\mathbf{v}_1$ and $\mathbf{v}_2$.

263. (A) Find $T(\mathbf{v}_2)$.
(B) Write $T(\mathbf{v}_2)$ as a linear combination of $\mathbf{v}_1$ and $\mathbf{v}_2$.

264. (A) Find $[T]_B$, the matrix representation of T in the basis B.
(B) Find $[T]_E$, the matrix representation of S in the usual basis
$E = \{\mathbf{e}_1 = (1, 0), \mathbf{e}_2 = (0, 1)\}$.

265. Let $A = \begin{bmatrix} 1 & 2 \\ 3 & 4 \end{bmatrix}$ and let $T: \mathbf{R}^2 \to \mathbf{R}^2$ be defined by $T(\mathbf{v}) = A\mathbf{v}$. Find $[T]_B$ relative to the above basis $B = \{\mathbf{v}_1 = (1, 3), \mathbf{v}_2 = (2, 5)\}$.

Theorem 8.2: Let the n-square matrix A be viewed as a linear operator T on K^n defined by $T(\mathbf{v}) = A\mathbf{v}$. Then the matrix representation of T relative to the usual basis of K^n is A itself.

266. Let $A = \begin{bmatrix} 1 & 2 \\ 3 & 4 \end{bmatrix}$ and let $T: \mathbf{R}^2 \to \mathbf{R}^2$ be defined by $T(\mathbf{v}) = A\mathbf{v}$. Find $[T]_E$ relative to the usual basis E.

267. Let $A = \begin{bmatrix} 1 & 2 & 3 \\ 4 & 5 & 6 \\ 7 & 8 & 9 \end{bmatrix}$ and let $T: \mathbf{R}^3 \to \mathbf{R}^3$ be defined by $T(\mathbf{v}) = A\mathbf{v}$. Find $[T]_E$ relative to the usual basis E.

Matrices and Linear Operators on $\mathbf{R}^3$

Remark 8.2: The matrix representation of $T: \mathbf{R}^3 \to \mathbf{R}^3$ defined by

$$T(\mathbf{v}) = (a_1\mathbf{x} + a_2\mathbf{y} + a_3\mathbf{z},\ b_1\mathbf{x} + b_2\mathbf{y} + b_3\mathbf{z},\ c_1\mathbf{x} + c_2\mathbf{y} + c_3\mathbf{z})$$

in the usual basis E is the matrix whose rows are the coefficients of x, y, z in $T(x, y, z)$: that is,

$$[T]_E = \begin{bmatrix} a_1 & a_2 & a_3 \\ b_1 & b_2 & b_3 \\ c_1 & c_2 & c_3 \end{bmatrix}.$$

This holds analogously for any linear map $T: \mathbf{R}^n \to \mathbf{R}^n$.

Problems 268–270 refer to $S: \mathbf{R}^3 \to \mathbf{R}^3$ defined by $S(x, y, z) = (x + 2y - 3z, 2x + y + z, 5x - y)$ and to the following basis B of $\mathbf{R}^3$:

$$B = \{\mathbf{u}_1 = (1, 1, 0), \mathbf{u}_2 = (1, 2, 3), \mathbf{u}_3 = (1, 3, 5)\}.$$

268. Find the coordinates of (a, b, c) with respect to the basis B.

269. (A) Find $S(\mathbf{u}_1)$.
(B) Write $S(\mathbf{u}_1)$ as a linear combination of $\mathbf{u}_1, \mathbf{u}_2, \mathbf{u}_3$.

270. (A) Find $S(\mathbf{u}_2)$.

(B Write $S(\mathbf{u}_2)$ as a linear combination of $\mathbf{u}_1$, $\mathbf{u}_2$, $\mathbf{u}_3$.

271. (A) Find $S(\mathbf{u}_3)$.

(B) Write $S(\mathbf{u}_3)$ as a linear combination of $\mathbf{u}_1$, $\mathbf{u}_2$, $\mathbf{u}_3$.

272. Find $[S]_B$, the matrix representation of S in the basis B.

273. Find $[S]_E$, the matrix representation of S in the usual basis E.

Questions 274–276 refer to $\mathbf{v} = (1, 1, 1)$ and the map S: $\mathbf{R}^3 \to \mathbf{R}^3$ and basis B in Problem 268.

274. Find $[\mathbf{v}]$.

275. Find:

(A) $S(\mathbf{v})$

(B) $[S(\mathbf{v})]$

276. Verify Theorem 8.1 that $[S][\mathbf{v}] = [S(\mathbf{v})]$.

Questions 277–283 refer to T: $\mathbf{R}^3 \to \mathbf{R}^3$ defined by $T(x, y, z) = (2y + z, x - 4y, 3z)$ and to the following basis B of $\mathbf{R}^3$:

$$B = \{\mathbf{w}_1 = (1, 1, 1), \mathbf{w}_2 = (1, 1, 0), \mathbf{w}_3 = (1, 0, 0)\}.$$

277. Find the coordinates of an arbitrary vector (a, b, c) with respect to the basis B.

278. (A) Find $T(\mathbf{w}_1)$.

(B) Write $T(\mathbf{w}_1)$ as a linear combination of $\mathbf{w}_1$, $\mathbf{w}_2$, $\mathbf{w}_3$.

279. (A) Find $T(\mathbf{w}_2)$.

(B) Write $T(\mathbf{w}_2)$ as a linear combination of $\mathbf{w}_1$, $\mathbf{w}_2$, $\mathbf{w}_3$.

280. (A) Find $T(\mathbf{w}_3)$.

(B) Write $T(\mathbf{w}_3)$ as a linear combination of $\mathbf{w}_1$, $\mathbf{w}_2$, $\mathbf{w}_3$.

281. (A) Find $[T]_B$, the matrix representation of S in the basis B.

(B) Find $[T]_E$, the matrix representation of S in the usual basis E.

282. Write $T(\mathbf{v})$ as a linear combination of $\mathbf{w}_1, \mathbf{w}_2, \mathbf{w}_3$ where $\mathbf{v} = (a, b, c)$.

283. Verify Theorem 8.1 that $[T][\mathbf{v}] = [T(\mathbf{v})]$ where $\mathbf{v} = (a, b, c)$.

Matrices and Linear Mapping Operations

Theorem 8.3: Let B be a basis of V over K, and let $\mathbf{M}$ be the algebra of n-square matrices over K. Let $m\colon A(V) \to \mathbf{M}$ be defined by $m(T) = [T]_B$. Then

(i) $m(T + S) = m(T) + m(S)$; that is, $[T + S] = [T] + [S]$.
(ii) $m(kT) = km(T)$; that is, $[kT] = k[T]$.
(iii) $m(S \circ T) = m(S)m(T)$; that is, $[S \circ T] = [S][T]$.
(iv) The mapping m is one-to-one and onto.

Questions 284–291 refer to the basis $B = \{\mathbf{u}_1 = (1, 1), \mathbf{u}_2 = (1, 2)\}$ of $\mathbf{R}^2$ and to the maps S and T defined by $S(x, y) = (x + 2y, 4x)$ and $T(x, y) = (y, x + 3y)$.

284. Find the coordinates of (a, b) in the basis B. [The formula for (a, b) will be used repeatedly.]

285. (A) Find $[S]$ in the basis B.
(B) Find $[T]$ in the basis B.

286. Write, as a linear combination of $\mathbf{u}_1$ and $\mathbf{u}_2$,
(A) $(S + T)(\mathbf{u}_1)$
(B) $(S + T)(\mathbf{u}_2)$

287. (A) Find $[S + T]$.
(B) Verify Theorem 8.3(i): $[S] + [T] = [S + T]$.

288. Write, as a linear combination of $\mathbf{u}_1$ and $\mathbf{u}_2$,
(A) $(3T)(\mathbf{u}_1)$
(B) $(3T)(\mathbf{u}_2)$

289. (A) Find $[3T]$.
(B) Verify Theorem 8.3(ii): $3[T] = [3T]$.

290. Write, as a linear combination of $\mathbf{u}_1$ and $\mathbf{u}_2$,
(A) $(S \circ T)(\mathbf{u}_1)$
(B) $(S \circ T)(\mathbf{u}_2)$

291. (A) Find $[S \circ T]$.
   (B) Verify Theorem 8.3(iii): $[S][T] = [S \circ T]$.

292. Prove Theorem 8.3(iv): The mapping m defined by $m(T) = [T]$ is one-to-one and onto.

Matrices and Linear Mapping from One Vector Space into Another

Here we consider the general case of linear mappings from one space into another.

Definition: Let $F: V \rightarrow U$ be linear. Let $\{\mathbf{e}_1, \ldots, \mathbf{e}_m\}$ be a basis of V and let $\{\mathbf{f}_1, \ldots, \mathbf{f}_n\}$ be a basis of U. Each $f(\mathbf{e}_i)$ is a linear combination of the basis elements $\mathbf{f}_j$ of U, say

$$f(\mathbf{e}_i) = \mathbf{a}_{i1}\mathbf{f}_1 + \mathbf{a}_{i2}\mathbf{f}_2 + \cdots + \mathbf{a}_{in}\mathbf{f}_n.$$

Then the transpose of the matrix of coefficients, denoted by $[F]_e^f$ or simply $[F]$, is called the *matrix representation of F in the basis* $\{\mathbf{e}_i\}$ *and* $\{\mathbf{f}_j\}$. [Note that [F] is an $n \times m$ matrix where $\dim(V) = m$ and $\dim(U) = n$.]

Theorem 8.4: For any vector $\mathbf{v}$ in V, $[F]_e^f [v]_e = [\mathbf{v}]_f$.

Questions 293–297 refer to $F: \mathbf{R}^3 \rightarrow \mathbf{R}^2$ defined by $F(x, y, z) = (2x + 3y - z, 4x - y + 2z)$ and to the following bases of $\mathbf{R}^3$ and $\mathbf{R}^2$, respectively:

$B_1 = \{\mathbf{u}_1 = (1, 1, 0), \mathbf{u}_2 = (1, 2, 3), \mathbf{u}_3 = (1, 3, 5)\}$ and $B_2 = \{\mathbf{v}_1 = (1, 2), \mathbf{v}_2 = (2, 3)\}$.

293. Find the coordinates of (a, b) relative to the basis vectors $\mathbf{v}_1$, and $\mathbf{v}_2$.

294. Write $F(\mathbf{u}_1)$ as a linear combination of the basis vectors $\mathbf{v}_1$ and $\mathbf{v}_2$.

295. Write $F(\mathbf{u}_2)$ as a linear combination of the basis vectors $\mathbf{v}_1$ and $\mathbf{v}_2$.

296. Write $F(\mathbf{u}_3)$ as a linear combination of the basis vectors $\mathbf{v}_1$ and $\mathbf{v}_2$.

297. Find $[F]$.

Questions 298–300 refer to $\mathbf{v} = (2, 5, -3)$ and the map F and bases B_1 and B_2 in Problem 293.

298. Find $[v]_{B_1}$.

299. Find

    (A) $F(\mathbf{v})$

    (B) $[F(\mathbf{v})]_{B_2}$

300. Verify Theorem 8.4 that $[F][\mathbf{v}]_{B_1} = [F(\mathbf{v})]_{B_2}$.

Change of Basis and Linear Operators

Theorem 8.5: Let P be the change-of-basis matrix from a basis S_1 to a basis S_2 in a vector space V. Then for any linear operator T on V, $[T]_{S_2} = P^{-1}[T]_{S_1} P$.

Theorem 8.6: The change-of-basis matrix P from the usual basis E of K^n to a basis $S = \{\mathbf{u}_1, \ldots, \mathbf{u}_n\}$ is the matrix P whose columns are the vectors $\mathbf{u}_1, \ldots, \mathbf{u}_n$.

Theorem 8.7: Let A be an n-square matrix over k (which may be viewed as a linear operator on K^n) and let $S = \{\mathbf{u}_1, \mathbf{u}_2, \ldots, \mathbf{u}_n\}$ be a basis of K^n. Then $B = P^{-1}AP$ is the matrix representation of A in the basis S, where P is the matrix whose columns are $\mathbf{u}_1, \mathbf{u}_2, \ldots, \mathbf{u}_n$, respectively.

Questions 301–304 refer to $T: \mathbf{R}^2 \rightarrow \mathbf{R}^2$ defined by $T(x, y) = (3x - 5y, 2x + 7y)$ and the basis $S = \{\mathbf{v}_1 = (1, 3), \mathbf{v}_2 = (2, 5)\}$ of $\mathbf{R}^2$.

301. Find the matrix representation of T relative to the usual basis E of $\mathbf{R}^2$.

302. Find the matrix representation of T relative to the basis S.

303. Find the change-of-basis matrix P from the usual basis E to S, and find P^{-1}.

304. Verify Theorem 8.5 that $[T]_S = P^{-1}[T]_E P$.

305. Let $L: \mathbf{R}^2 \rightarrow \mathbf{R}^2$ be defined by $L(x, y) = (2y, 3x - y)$. Find $[L]_S$ for the basis S in Problem 302.

306. Let $A: \mathbf{R}^2 \rightarrow \mathbf{R}^2$ be defined by the matrix $A = \begin{bmatrix} 5 & -7 \\ 2 & 3 \end{bmatrix}$. Let B be the matrix representation of A relative to the basis $\{(1, 4), (3, 10)\}$. Find B.

307. Given $P = \begin{bmatrix} 1 & 2 & 1 \\ 1 & 3 & 4 \\ 2 & 5 & 6 \end{bmatrix}$ and $P^{-1} = \begin{bmatrix} -2 & -7 & 5 \\ 2 & 4 & -3 \\ -1 & -1 & 1 \end{bmatrix}$, let $A = \begin{bmatrix} 2 & 3 & -4 \\ 4 & -6 & 3 \\ 1 & 4 & -2 \end{bmatrix}$

and let B be the matrix representation of $A: \mathbf{R}^3 \rightarrow \mathbf{R}^3$ in the basis $S = \{(1, 1, 2), (2, 3, 5), (1, 4, 6)\}$. Find B.

308. Given $P = \begin{bmatrix} 1 & 1 & 1 \\ 0 & 1 & 1 \\ 0 & 0 & 1 \end{bmatrix}$ and $P^{-1} = \begin{bmatrix} 1 & -1 & 0 \\ 0 & 1 & -1 \\ 0 & 0 & 1 \end{bmatrix}$, let $A = \begin{bmatrix} 1 & 3 & 5 \\ 2 & 4 & 6 \\ 7 & 8 & 9 \end{bmatrix}$ and let B

be the matrix representation of $A: \mathbf{R}^3 \to \mathbf{R}^3$ in the basis $S = \{(1, 0, 0), (1, 1, 0), (1, 1, 1)\}$. Find B.

Similarity

Definition: Let A and B be square matrices for which there exists an invertible matrix P such that $B = P^{-1} AP$. Then B is said to be similar to A.

Questions 309–311 show that similarity of matrices is an equivalence relation.

309. Show that A is similar to A for any (square) matrix A.

310. Suppose A is similar to B. Show that B is similar to A.

311. Suppose A is similar to B and B is similar to C. Show that A is similar to C.

Theorem 8.8: Suppose matrix B is similar to matrix A. Then
(i) $\text{tr}(B) = \text{tr}(A)$ and
(ii) $\det(B) = \det(A)$, where $\text{tr}(A)$ is the trace of A and $\det(A)$ is the determinant of A.

312. Let $F: \mathbf{R}^2 \to \mathbf{R}^2$ be defined by $F(x, y) = (3x - 7y, 4x + 8y)$. Let A be the matrix representation of F relative to the usual basis E. Find A.

313. For the map F in Question 312, find:
  (A) $\text{tr}(F)$
  (B) $\det(F)$

314. Let $T: \mathbf{R}^3 \to \mathbf{R}^3$ be defined by $T(x, y, z) = (2x - z, x + 2y - 4z, 3x - 3y + z)$. Let B be the matrix representation of T relative to the usual basis E. Find B.

315. For the map T in Question 314, find:
  (A) $\text{tr}(T)$
  (B) $\det(T)$

CHAPTER 9

Inner Product Spaces

The definition of a vector space V involves a field K of scalars. Here we restrict K to be either the real field **R** or the complex field **C**. First we assume $K = \mathbf{R}$, in which case V is called a *real vector space*. In the last sections, we extend our results to the case $K = \mathbf{C}$, in which case V is called a *complex vector space*.

Inner Product Spaces

Definition: Let V be a real vector space. Suppose to each pair of vectors $\mathbf{u}, \mathbf{v} \in V$ there is assigned a real number, denoted by $\langle \mathbf{u}, \mathbf{v} \rangle$. This function is called a (real) *inner product* on V if it satisfies the follow axioms where $\mathbf{u}_1, \mathbf{u}_2, \mathbf{u}, \mathbf{v} \in V$ and $a, b, k \in \mathrm{R}$:

[RIP$_1$: Linear Property] $\langle a\mathbf{u}_1 + b\mathbf{u}_2, \mathbf{v} \rangle = a\langle \mathbf{u}_1, \mathbf{v} \rangle + b\langle \mathbf{u}_2, \mathbf{v} \rangle$ or, equivalently,

  (i) $\langle \mathbf{u}_1 + \mathbf{u}_2, \mathbf{v} \rangle = \langle \mathbf{u}_1, \mathbf{v} \rangle + \langle \mathbf{u}_2, \mathbf{v} \rangle$ and
  (ii) $\langle k\mathbf{u}, \mathbf{v} \rangle = k\langle \mathbf{u}, \mathbf{v} \rangle$.

[RIP$_2$: Symmetric Property] $\langle \mathbf{u}, \mathbf{v} \rangle = \langle \mathbf{v}, \mathbf{u} \rangle$.
[RIP$_3$: Positive Definite Property] If $\mathbf{u} \neq \mathbf{0}$, then $\langle \mathbf{u}, \mathbf{u} \rangle > 0$.

The vector space V with an inner product is called an *inner product space* or, simply, *IPS*.

Definition: Let V be an IPS. The *norm* or *length* of $\mathbf{u} \in V$ is denoted and defined by

$$\|\mathbf{u}\| = \sqrt{\langle \mathbf{u}, \mathbf{u} \rangle}.$$

(The relation $\|\mathbf{u}\|^2 = \langle \mathbf{u}, \mathbf{u} \rangle$ will be frequently used.) The *distance* between $\mathbf{u}, \mathbf{v} \in V$ is denoted and defined by $d(\mathbf{u}, \mathbf{v}) = \|\mathbf{u} - \mathbf{v}\|$.

Definition: A vector $\mathbf{u}$ is a *unit vector* if $\|\mathbf{u}\| = 1$. For vector $\mathbf{v} \neq \mathbf{0}$, the vector $\hat{\mathbf{v}} = (1/\|\mathbf{v}\|)\,\mathbf{v} = \mathbf{v}/\|\mathbf{v}\|$ is the unique unit vector in the direction of $\mathbf{v}$. The process of finding $\hat{\mathbf{v}}$ from $\mathbf{v}$ is called *normalizing* $\mathbf{v}$.

Definition: Let $\mathbf{u} = (a_1, a_2, \ldots, a_n)$ and $\mathbf{v} = (b_1, b_2, \ldots, b_n)$ be vectors in $\mathbf{R}^n$. The *dot product* of $\mathbf{u}$ and $\mathbf{v}$, denoted and defined by

$$\mathbf{u} \cdot \mathbf{v} = a_1 b_1 + a_2 b_2 + \cdots + a_n b_n,$$

is an inner product on $\mathbf{R}^n$ called the *usual* or *standard inner product* on $\mathbf{R}^n$. We assume this inner product on $\mathbf{R}^n$ unless otherwise stated or implied. $\mathbf{R}^n$ with this inner product is called *Euclidean n*-space.

Remark 9.1: Assuming $\mathbf{u}$ and $\mathbf{v}$ are column vectors, the usual inner product may be defined by $\langle \mathbf{u}, \mathbf{v} \rangle = \mathbf{u}^T \mathbf{v}$.

316. Show that $\langle \mathbf{0}, \mathbf{v} \rangle = 0 = \langle \mathbf{v}, \mathbf{0} \rangle$ for every $\mathbf{v}$ in V.

317. Show that $\langle \mathbf{u}, \mathbf{v}_1 + \mathbf{v}_2 \rangle = \langle \mathbf{u}, \mathbf{v}_1 \rangle + \langle \mathbf{u}, \mathbf{v}_2 \rangle$.

318. Show that $\langle \mathbf{u}, k\mathbf{v} \rangle = k\langle \mathbf{u}, \mathbf{v} \rangle$.

Problems 319–321 refer to the vectors $\mathbf{u} = (1, 2, 4)$, $\mathbf{v} = (2, -3, 5)$, $\mathbf{w} = (4, 2, -3)$ in $\mathbf{R}^3$.

319. Find:
   (A) $\mathbf{u} \cdot \mathbf{v}$
   (B) $\mathbf{u} \cdot \mathbf{w}$
   (C) $\mathbf{v} \cdot \mathbf{w}$

320. Find $(\mathbf{u} + \mathbf{v}) \cdot \mathbf{w}$.

321. Find:
   (A) $\|\mathbf{u}\|$
   (B) $\|\mathbf{v}\|$
   (C) $\|\mathbf{u} + \mathbf{v}\|$

322. Let V be the vector space of polynomials with inner product defined by

$$\langle \mathbf{f}, \mathbf{g} \rangle = \int_0^1 f(t)g(t)\, dt.$$

Let $f(t) = t + 2$, $g(t) = 3t - 2$, $h(t) = t^2 - 2t - 3$. Find:
(A) $\langle \mathbf{f}, \mathbf{g} \rangle$
(B) $\langle \mathbf{f}, \mathbf{h} \rangle$

323. For polynomials $f(t)$ and $g(t)$ in Question 322, find:

(A) $||\mathbf{f}||$

(B) $||\mathbf{g}||$

324. Show that $||\mathbf{u} + \mathbf{v}||^2 = ||\mathbf{u}||^2 + 2\langle \mathbf{u}, \mathbf{v} \rangle + ||\mathbf{v}||^2$.

325. Show that $||\mathbf{u} - \mathbf{v}||^2 = ||\mathbf{u}||^2 - 2\langle \mathbf{u}, \mathbf{v} \rangle + ||\mathbf{v}||^2$.

326. Verify the following polar form: $\langle \mathbf{u}, \mathbf{v} \rangle = \frac{1}{4} (||\mathbf{u} + \mathbf{v}|| - ||\mathbf{u} - \mathbf{v}||)$ (which shows the inner product can be obtained from the norm function).

327. For $\mathbf{v} \neq \mathbf{0}$, show that vector $\hat{\mathbf{v}} = (1/||\mathbf{v}||)/\mathbf{v} = \mathbf{v}/||\mathbf{v}||$ is the unique unit vector in the direction of $\mathbf{v}$.

328. Normalize:

(A) $\mathbf{u} = (2, 1, -1)$ in $\mathbf{R}^3$

(B) $\mathbf{v} = (\frac{1}{2}, \frac{1}{3}, -\frac{1}{4})$ in $\mathbf{R}^3$

329. Find $d(\mathbf{u}, \mathbf{v})$ where $\mathbf{u} = (5, 4)$, $\mathbf{v} = (2, -6)$ in $\mathbf{R}^2$.

330. Find $d(\mathbf{u}, \mathbf{v})$ where $\mathbf{u} = (5, 5, 8, 8)$, $\mathbf{v} = (1, 2, 3, 4)$ in $\mathbf{R}^4$.

331. Find $d(\mathbf{f}, \mathbf{g})$ for the polynomials $f(t)$ and $g(t)$ in Problem 322.

Theorem 9.1 (Cauchy–Schwarz Inequality): For any vectors $\mathbf{u}, \mathbf{v} \in V$, $\langle \mathbf{u}, \mathbf{v} \rangle^2 \leq ||\mathbf{u}||^2||\mathbf{v}||^2$, or $|\langle \mathbf{u}, \mathbf{v} \rangle| \leq ||\mathbf{u}|| \, ||\mathbf{v}||$.

Remark 9.2: The angle θ between nonzero vectors $\mathbf{u}, \mathbf{v} \in V$ is the angle $0 \leq \theta \leq \pi$ where $\cos \theta = \langle \mathbf{u}, \mathbf{v} \rangle / ||\mathbf{u}|| \, ||\mathbf{v}||$.

332. Find $\cos \theta$ for angle θ between

(A) $\mathbf{u} = (1, -3, 2)$, and $\mathbf{v} = (2, 1, 5)$ in $\mathbf{R}^3$

(B) $\mathbf{u} = (5, 1)$, and $\mathbf{v} = (-2, 3)$ in $\mathbf{R}^2$

333. Find $\cos \theta$ for angle θ between $f(t) = 2t - 1$, $g(t) = t^2$ where

$$\langle \mathbf{f}, \mathbf{g} \rangle = \int_0^1 f(t)g(t) \, dt.$$

334. Show that the norm satisfies the following three properties:

(A) $\mathbf{N_1}$: $||\mathbf{v}|| \geq 0$ and $||\mathbf{v}|| = 0$ iff $\mathbf{v} = \mathbf{0}$

(B) $\mathbf{N_2}$: $||k\mathbf{v}|| = |k| \, ||\mathbf{v}||$

(C) $\mathbf{N_3}$: $||\mathbf{u} + \mathbf{v}|| \leq ||\mathbf{u}|| + ||\mathbf{v}||$

Orthogonality, Orthogonal Complements

Definition: Let V be an IPS. The vectors $\mathbf{u}, \mathbf{v} \in V$ are orthogonal if $\langle \mathbf{u}, \mathbf{v} \rangle = 0$. If W is a subspace of V, then the *orthogonal complement* of W, written $W^\perp$ (read "W perp"), consists of those vectors in V that are orthogonal to every vector $\mathbf{w} \in W$.

335. Show that $W^\perp$ is a subspace of V.

336. Let $\mathbf{u} = (1, 3, -4)$ in $\mathbf{R}^3$. Find a basis for $\mathbf{u}^\perp$.

337. Let W consist of the vectors $\mathbf{u} = (1, 2, 3, -1, 2)$ and $\mathbf{v} = (2, 4, 7, 2, -1)$ in $\mathbf{R}^5$. Find a basis for $W^\perp$.

338. Consider a homogeneous system $AX = 0$. Recall that the solution space W may be viewed as the kernel of the linear mapping A. Give another interpretation of W using the notion of orthogonality.

Definition: Consider a set $S = \{\mathbf{u}_1, \mathbf{u}_2, \ldots, \mathbf{u}_k\}$ of vectors in an inner product space V. S is said to be *orthogonal* if each of its vectors is nonzero and if its vectors are mutually orthogonal, i.e., if $\langle \mathbf{u}_i, \mathbf{u}_i \rangle \neq 0$ but $\langle \mathbf{u}_i, \mathbf{u}_j \rangle = 0$ for $i \neq j$. S is said to be *orthonormal* if S is orthogonal and if each of its vectors has unit length or, in other words, if

$$\langle \mathbf{u}_i, \mathbf{u}_j \rangle = \delta_{ij} = \begin{cases} 1 & \text{if } i = j \\ 0 & \text{if } i \neq j \end{cases}.$$

Normalizing refers to the process of dividing each vector in an orthogonal set S by its length so S is transformed into an orthonormal set. An orthogonal (orthonormal) basis refers to a basis S that is also orthogonal (orthonormal).

339. Show that $S = \{\mathbf{u} = (1, 2, -3, 4), \mathbf{v} = (3, 4, 1, -2), \mathbf{w} = (3, -2, 1, 1)\}$ is orthogonal.

340. Normalize the orthogonal set in Problem 339 to obtain an orthonormal set.

341. Show that the usual basis E in $\mathbf{R}^3$ is orthonormal.

Theorem 9.2: An orthogonal set S of vectors is linearly independent.

Questions 342–344 refer to the set of vectors

$$S = \{(\mathbf{u}_1 = (1, 2, 1), \mathbf{u}_2 = (2, 1, -4), \mathbf{u}_3 = (3, -2, 1)\}.$$

342. (A) Show that S is orthogonal.
(B) Is S a basis of $\mathbf{R}^3$?

343. Write $\mathbf{w} = (3, 4, 5)$ as a linear combination of $\mathbf{u}_1$, $\mathbf{u}_2$, $\mathbf{u}_3$. (Hint: Use that S is orthogonal.)

344. Normalize S to get an orthonormal basis of $\mathbf{R}^3$.

Theorem 9.3: Suppose $\{\mathbf{u}_1, \mathbf{u}_2, \ldots, \mathbf{u}_n\}$ is an orthogonal basis for V. Then, for any $\mathbf{v} \in V$,

$$\mathbf{v} = \frac{\langle \mathbf{v}, \mathbf{u}_1 \rangle}{\langle \mathbf{u}_1, \mathbf{u}_1 \rangle} \mathbf{u}_1 + \frac{\langle \mathbf{v}, \mathbf{u}_2 \rangle}{\langle \mathbf{u}_2, \mathbf{u}_2 \rangle} \mathbf{u}_2 + \cdots + \frac{\langle \mathbf{v}, \mathbf{u}_n \rangle}{\langle \mathbf{u}_n, \mathbf{u}_n \rangle} \mathbf{u}_n.$$

Remark 9.3: The above scalar $k = \langle \mathbf{v}, \mathbf{u}_i \rangle / \|\mathbf{u}_i\|^2$ is called the *component* of $\mathbf{v}$ along $\mathbf{u}_i$ or the *Fourier coefficient* of $\mathbf{v}$ with respect to $\mathbf{u}_i$.

345. Let $\mathbf{w} = (1, 2, 3)$ in $\mathbf{R}^3$. Find a basis of the orthogonal complement $\mathbf{w}^\perp$ of $\mathbf{w}$.

346. Find an orthonormal basis of the orthogonal complement $\mathbf{w}^\perp$ of $\mathbf{w}$ in Problem 345.

347. Show that if $\{\mathbf{u}_1, \mathbf{u}_2, \ldots, \mathbf{u}_r\}$ is orthogonal, then $\{a_1\mathbf{u}_1, a_2\mathbf{u}_2, \ldots, a_r\mathbf{u}_r\}$ is orthogonal for any choice of nonzero scalars $a_1, a_2, \ldots, a_r$.

Orthogonal Matrices

Definition: A real matrix A is orthogonal if $A^{-1} = A^T$,—that is, if $AA^T = A^TA = I$.

Theorem 9.4: A square matrix P is orthogonal if and only if the rows (columns) of P form an orthonormal set.

Theorem 9.5: Any 2-square orthogonal matrix has the form $\begin{bmatrix} \cos\theta & \sin\theta \\ -\sin\theta & \cos\theta \end{bmatrix}$ or $\begin{bmatrix} \cos\theta & \sin\theta \\ \sin\theta & -\cos\theta \end{bmatrix}$ for some real number θ.

348. Find an orthogonal matrix P with first row $(1/\sqrt{10}, 3/\sqrt{10})$.

349. Find an orthogonal matrix P with first row $(1/3, 2/3, 2/3)$.

Projections, Gram–Schmidt Algorithm

350. Suppose $\mathbf{w} \neq \mathbf{0}$. Let $\mathbf{v}$ be any vector. Show that $c = \langle \mathbf{u}, \mathbf{w} \rangle / \langle \mathbf{w}, \mathbf{w} \rangle = \langle \mathbf{u}, \mathbf{w} \rangle / \|\mathbf{w}\|^2$ is the unique scalar such that $\mathbf{v}' = c - c\mathbf{w}$ is orthogonal to $\mathbf{w}$.

Remark 9.4: The scalar c is called the Fourier coefficient of $\mathbf{v}$ with respect to $\mathbf{w}$. The vector $c\mathbf{w}$ is called the *projection* of $\mathbf{v}$ along $\mathbf{w}$.

351. Find the Fourier coefficient c and the projection $c\mathbf{w}$ of $\mathbf{v} = (1, -1, 2)$ along $\mathbf{w} = (0, 1, 1)$.

Algorithm 9.1 (Gram–Schmidt): The input is a basis $\mathbf{v}_1, \mathbf{v}_2, \ldots, \mathbf{v}_r$ of a subspace U of V. The output is an orthogonal basis of U.
Set

$$\mathbf{w}_1 = \mathbf{v}_1$$

$$\mathbf{w}_2 = \mathbf{v}_2 - c_{21}\mathbf{w}_1 = \mathbf{v}_2 - \frac{\langle \mathbf{v}_2, \mathbf{w}_1 \rangle}{\|\mathbf{w}_1\|^2}\mathbf{w}_1$$

$$\mathbf{w}_3 = \mathbf{v}_3 - c_{31}\mathbf{w}_1 - c_{32}\mathbf{w}_2 = \mathbf{v}_3 - \frac{\langle \mathbf{v}_3, \mathbf{w}_1 \rangle}{\|\mathbf{w}_1\|^2}\mathbf{w}_1 - \frac{\langle \mathbf{v}_3, \mathbf{w}_2 \rangle}{\|\mathbf{w}_2\|^2}\mathbf{w}_2$$

$$\mathbf{w}_r = \mathbf{v}_r - c_{r1}\mathbf{w}_1 - c_{r2}\mathbf{w}_2 - \cdots - c_r, \mathbf{w}_{r-1}$$

where $c_{ri} = \langle \mathbf{v}_r, \mathbf{w}_i \rangle / \|\mathbf{w}_i\|^2$. The set $\{\mathbf{w}_1, \mathbf{w}_2, \ldots, \mathbf{w}_r\}$ is the required orthogonal basis of U.

Remark 9.5: In hand calculations, it may be simpler to clear fractions in any new $\mathbf{w}_k$.

352. Find an orthonormal basis for the subspace U of $\mathbf{R}^4$ spanned by

$$\mathbf{v}_1 = (1, 1, 1, 1), \mathbf{v}_2 = (1, 2, 4, 5), \mathbf{v}_3 = (1, -3, -4, -2).$$

353. Consider the following basis of $\mathbf{R}^3$: $S = \{\mathbf{v}_1 = (1, 1, 1), \mathbf{v}_2 = (0, 1, 1), \mathbf{v}_3 = (0, 0, 1)\}$. Use the Gram–Schmidt algorithm to transform $\{\mathbf{v}_i\}$ into an orthonormal basis $\{\mathbf{u}_i\}$ of $\mathbf{R}^3$.

Theorem 9.6 (Bessel Inequality): Suppose $\{\mathbf{u}_1, \mathbf{u}_2, \ldots, \mathbf{u}_r\}$ is an orthonormal set of vectors in V. Let $\mathbf{v} \in V$, and let c_i be the Fourier coefficient of $\mathbf{v}$ with respect to $\mathbf{u}_i$. Then

$$\sum_i c_i^2 \leq \|\mathbf{v}\|^2.$$

354. Prove Theorem 9.6, the Bessel inequality.

Inner Products and Positive Definite Matrices

Definition: Let $B = \{e_1, e_2, \ldots, e_n\}$ be a basis of an IPS V. The matrix $A = [\langle e_i, e_j \rangle]$ is said to represent the inner product on V with respect to the basis B.

355. Consider the basis $B = \{u_1 = (1, 1, 0), u_2 = (1, 2, 3), u_3 = (1, 3, 5)\}$ of $\mathbf{R}^3$. Find the matrix A that represents the usual inner product on $\mathbf{R}^3$ with respect to the basis B.

356. Consider the usual basis $E = \{e_1 = (1, 0, 0), e_2 = (0, 1, 0), e_3 = (0, 0, 1)\}$ of $\mathbf{R}^3$. Find the matrix that represents the usual inner product on $\mathbf{R}^3$ with respect to the usual basis E.

Theorem 9.7: Let A be the matrix representing an inner product on V with respect to a basis $B = \{e_1, \ldots, e_n\}$. Then, for any vectors $\mathbf{u}, \mathbf{v} \in V$, $\langle \mathbf{u}, \mathbf{v} \rangle = [\mathbf{u}]^T A[\mathbf{v}]$ where $[\mathbf{u}]$ and $[\mathbf{v}]$ denote, respectively, the (column) coordinate vectors of $\mathbf{u}$ and $\mathbf{v}$ relative to the basis B.

Questions 357–359 refer to the vector space V of polynomials of degree ≤ 2 with inner product defined by $\langle \mathbf{f}, \mathbf{g} \rangle = \int_{-1}^{1} f(t) \, g(t) \, dt$. Let $f(t) = t + 2$ and $g(t) = t^2 - 3t + 4$.

357. Find $\langle \mathbf{f}, \mathbf{g} \rangle$.

358. Find the matrix A of the inner product with respect to the basis $\{1, t, t^2\}$ of V.

359. Verify Theorem 9.7 that $\langle \mathbf{f}, \mathbf{g} \rangle = [\mathbf{f}]^T A[\mathbf{g}]$ with respect to the basis $\{1, t, t^2\}$ of V.

Definition: A square matrix A is positive definite if A is symmetric and if $\mathbf{x}^T A\mathbf{x} > 0$ for every nonzero vector $\mathbf{x}$.

Theorem 9.8: Let A be a matrix that represents an inner product on V with respect to any basis B of V. Then A is positive definite.

360. Prove Theorem 9.8.

Complex Inner Product Spaces

Let V be a vector space over the complex field $\mathbf{C}$. Recall that, for $z = a + bi \in \mathbf{C}$,

$$\bar{z} = a - bi, \quad z\bar{z} = a^2 + b^2, \quad |z| = \sqrt{a^2 + b^2}, \text{ and } z \text{ is real if } z = \bar{z}.$$

Definition: Suppose to each pair of vectors $\mathbf{u}, \mathbf{v} \in V$ there is assigned a complex number, denoted by $\langle \mathbf{u}, \mathbf{v} \rangle$. Then this function $\langle, \rangle$ is called a *complex inner product* on V if it satisfies the following axioms (where $\mathbf{u}_1, \mathbf{u}_2, \mathbf{u}, \mathbf{v} \in V$ and $a, b, k \in \mathbf{C}$).

[CIP$_1$: Linear Property] $\langle a\mathbf{u}_1 + b\mathbf{u}_2, \mathbf{v} \rangle = a\langle \mathbf{u}_1, \mathbf{v} \rangle + b\langle \mathbf{u}_2, \mathbf{v} \rangle$ or, equivalently,
 (i) $\langle \mathbf{u}_1, + \mathbf{u}_2, \mathbf{v} \rangle = \langle \mathbf{u}_1, \mathbf{v} \rangle + \langle \mathbf{u}_2, \mathbf{v} \rangle$ and
(ii) $\langle k\mathbf{u}, \mathbf{v} \rangle = k\langle \mathbf{u}, \mathbf{v} \rangle$.

[CIP$_2$: Conjugate Symmetric Property] $\langle \mathbf{u}, \mathbf{v} \rangle = \overline{\langle \mathbf{v}, \mathbf{u} \rangle}$.
[CIP$_3$: (Positive Definite Property] If $\mathbf{u} \neq \mathbf{0}$, then $\langle \mathbf{u}, \mathbf{u} \rangle > 0$.

The complex vector space V with an inner product is called a complex inner product space.

361. Show that $\langle \mathbf{u}, k\mathbf{v} \rangle = \bar{k}\langle \mathbf{u}, \mathbf{v} \rangle$. That is, we must take the conjugate of a complex scalar when it is taken out of the second position of the inner product.

362. Suppose $\langle \mathbf{u}, \mathbf{v} \rangle = 3 + 2i$. Find:
  (A) $\langle (2 - 4i)\mathbf{u}, \mathbf{v} \rangle$.
  (B) $\langle \mathbf{u}, (4 + 3i)\mathbf{v} \rangle$.

363. Suppose $\langle \mathbf{u}, \mathbf{v} \rangle = 3 + 2i$. Find $\langle (3 - 6i)\mathbf{u}, (5 - 2i)\mathbf{v} \rangle$.

Questions 364–365 refer to vectors $u = (1 - i, 2 + 3i)$ and $\mathbf{v} = (2 - 5i, 3 - i)$ in $\mathbf{C}^2$.

364. Find:
  (A) $\langle \mathbf{u}, \mathbf{v} \rangle$
  (B) $\langle \mathbf{v}, \mathbf{u} \rangle$

365. Find:
  (A) $\|\mathbf{u}\|$
  (B) $\|\mathbf{v}\|$

Eigenvalues, Eigenvectors, Diagonalization

Here we study conditions when a matrix A is similar to a diagonal matrix. This theory involves the roots of a polynomial associated with A, so the underlying field K plays an important part.

Characteristic Polynomial, Cayley-Hamilton Theorem

Definition: Let $A = [a_{ij}]$ be an n-square matrix. The matrix $tI_n - A$, where t is an indeterminant, is the *characteristic matrix* of A. Its determinant $\det(tI - A)$, which is a polynomial in t, is the *characteristic polynomial* of A, and $\det(tI - A) = 0$ is the *characteristic equation* of A.

Remark 10.1: Let $A = [a_{ij}]$ be an *n*-square matrix and $\Delta_A(t)$ its characteristic polynomial then,

$$\Delta_A(t) \text{ is } t^n = -\text{ tr}(A)t^{n-1} + \ldots + (-1)^n|A|.$$

Theorem 10.1 (Cayley-Hamilton): Every matrix A is a zero of its characteristic polynomial.

366. Let $A = \begin{bmatrix} 1 & 2 \\ 3 & 2 \end{bmatrix}$.

    (A) Find the characteristic polynomial $\Delta(t)$ of A.
    (B) Verify that A is a root of $\Delta(t)$.

367. Find the characteristic polynomial $\Delta(t)$ of each matrix:

    (A) $A = \begin{bmatrix} -2 & -6 \\ 4 & 9 \end{bmatrix}$

    (B) $B = \begin{bmatrix} 4 & 5 \\ -3 & -7 \end{bmatrix}$

Remark 10.2: Let $A = [a_{ij}]$ be a 3-square matrix and $\Delta(t)$ its characteristic polynomial. Then

$$\Delta(t) = t^3 - \text{tr}(A)t^2 + (A_{11} + A_{22} + A_{33})t^2 - \det(A)$$

where A_{11}, A_{22}, A_{33} are the cofactors of the diagonal elements of A.

Questions 368–369 refer to the following matrices:

$$A = \begin{bmatrix} 1 & 2 & 3 \\ 5 & 4 & 1 \\ 2 & 7 & 2 \end{bmatrix}, \quad B = \begin{bmatrix} 1 & 1 & 2 \\ 0 & 3 & 2 \\ 1 & 3 & 9 \end{bmatrix}.$$

368. Find the characteristic polynomial $\Delta(t)$ of A.

369. Find the characteristic polynomial $\Delta(t)$ of B.

Remark 10.3: Let $A = [a_{ij}]$ be an upper triangular matrix. Then its characteristic polynomial is

$$\Delta(t) = (t - a_{11})(t - a_{22}) \cdots (t - a_{nn}).$$

Let M be an upper triangular block matrix with diagonal blocks $A_1, A_2, \ldots, A_n$. Then its characteristic polynomial $\Delta_M(t)$ is the product of the characteristic polynomials of its diagonal blocks; that is,

$$\Delta_M(t) = \Delta_{A_1}(t)\Delta_{A_2}(t) \ldots \Delta_{A_n}(t).$$

Questions 370–372 refer to the following matrices:

$$R = \begin{bmatrix} 1 & 2 & 3 & 4 \\ 0 & 2 & 8 & -6 \\ 0 & 0 & 3 & -5 \\ 0 & 0 & 0 & 4 \end{bmatrix}, \quad S = \begin{bmatrix} 2 & 5 & 7 & -9 \\ 1 & 4 & -6 & 4 \\ 0 & 0 & 6 & -5 \\ 0 & 0 & 2 & 3 \end{bmatrix}, \quad T = \begin{bmatrix} 5 & 8 & -1 & 0 \\ 0 & 3 & 6 & 7 \\ 0 & -3 & 5 & -5 \\ 0 & 0 & 0 & 7 \end{bmatrix}.$$

370. Find the characteristic polynomial $\Delta(t)$ of R.

371. Find the characteristic polynomial $\Delta(t)$ of S.

372. Find the characteristic polynomial $\Delta(t)$ of T.

Eigenvalues and Eigenvectors

Definition: Let $A = [a_{ij}]$ be an n-square matrix over a field K. A scalar $\lambda \in K$ is called an *eigenvalue* of A if there exists a nonzero column vector $\mathbf{v}$ for which $A\mathbf{v} = \lambda\mathbf{v}$. Any vector satisfying this relation is called an *eigenvector* of A belonging to λ. The set E_λ of all eigenvectors belonging to λ is a subspace of K^n called the *eigenspace* of λ.

Remark 10.4: There are analogous definitions for a linear mapping $T: V \to V$.

Theorem 10.2: Nonzero eigenvectors belonging to distinct eigenvalues are linearly independent.

373. Let $A = \begin{bmatrix} 1 & 2 \\ 3 & 2 \end{bmatrix}$.

    (A) Show that $\mathbf{v}_1 = \begin{bmatrix} 2 \\ 3 \end{bmatrix}$ is an eigenvector of A belonging to $\lambda_1 = 4$.

    (B) Show that $\mathbf{v}_2 = \begin{bmatrix} 1 \\ -1 \end{bmatrix}$ is an eigenvector of A belonging to $\lambda_2 = -1$.

Theorem 10.3: An n-square matrix A is similar to a diagonal matrix B if and only if A has n linearly independent eigenvectors. In this case the diagonal elements of B are the corresponding eigenvalues, and $B = P^{-1}AP$ where P is the matrix whose columns are the eigenvectors.

374. Verify Theorem 10.3 for the matrix $A = \begin{bmatrix} 1 & 2 \\ 3 & 2 \end{bmatrix}$ in Problem 373.

Theorem 10.4: Let A be an n-square matrix over a field K. A scalar λ is an *eigenvalue* of A if and only if λ is a root of the characteristic polynomial $\Delta(t)$ of A.

Computing Eigenvalues and Eigenvectors, Diagonalizing Matrices

Algorithm 10.1 (Diagonalization Algorithm): The input is an n-square matrix A.
Step 1. Find the characteristic polynomial $\Delta(t)$ of A.
Step 2. Find the roots of $\Delta(t)$ to obtain the eigenvalues of A.
Step 3. Repeat (A) and (B) for each eigenvalue λ of A:
    (A) Form $M = A - \lambda I$ by subtracting λ down the diagonal of A, or form $M' = \lambda I - A$ by substituting $t = \lambda$ in $tI - A$.
    (B) Find a basis for the solution space of the homogeneous system $MX = 0$. (These basis vectors are linearly independent eigenvectors of A belonging to λ.)

Step 4. Consider the collection $S = \{\mathbf{v}_1, \mathbf{v}_2, \ldots, \mathbf{v}_m\}$ of all eigenvectors obtained in Step 3:

(A) If $m \neq n$, then A is not diagonalizable.

(B) If $m = n$, let P be the matrix whose columns are the eigenvectors $\mathbf{v}_1$, $\mathbf{v}_2, \ldots, \mathbf{v}_n$. Then

$$D = P^{-1}AP = \mathrm{diag}(\lambda_1, \lambda_2, \ldots, \lambda_n)$$

where λ_i is the eigenvalue corresponding to the eigenvector $\mathbf{v}_i$.

375. Let $A = \begin{bmatrix} 2 & 2 \\ 1 & 3 \end{bmatrix}$.

(A) Find all eigenvalues and a maximum set S of linearly independent eigenvectors.

(B) Is A diagonalizable? If so, find P and diagonal matrix D such that $D = P^{-1}AP$.

376. Let $A = \begin{bmatrix} 1 & -1 \\ 2 & -1 \end{bmatrix}$.

(A) Find all eigenvalues and a maximum set S of linearly independent eigenvectors, assuming A is a real matrix.

(B) Is A diagonalizable? If so, find P such that $P^{-1}AP$ is diagonal.

377. Let A be the matrix in Problem 376.

(A) Find all eigenvalues and a maximum set S of linearly independent eigenvectors, assuming A is a complex matrix.

(B) Is A diagonalizable? If so, find P such that $P^{-1}AP$ is diagonal.

378. Let $B = \begin{bmatrix} 2 & 4 \\ 3 & 1 \end{bmatrix}$.

(A) Find all eigenvalues and a maximum set S of linearly independent eigenvectors.

(B) Is B diagonalizable? If so, find P and diagonal matrix D such that $D = P^{-1}BP$.

Problems 379–381 refer to the matrix $C = \begin{bmatrix} 4 & 1 & -1 \\ 2 & 5 & -2 \\ 1 & 1 & 2 \end{bmatrix}$.

379. (A) Find the characteristic polynomial $\Delta(t)$ of C.

(B) Find all eigenvalues of C.

380. Find a maximum set S of linearly independent eigenvectors of C.

381. Is C diagonalizable? If so, find P such that $P^{-1} AP$ is diagonal.

382. Find a 2×2 matrix A with eigenvalues $\lambda_1 = 2$ and $\lambda_2 = 3$ and corresponding eigenvectors $\mathbf{v}_1 = (1, 3)$ and $\mathbf{v}_2 = (1, 4)$.

383. Find a 2×2 matrix B with eigenvalues $\lambda_1 = -1$ and $\lambda_2 = 2$ and corresponding eigenvectors $\mathbf{v}_1 = (1, 2)$ and $\mathbf{v}_2 = (3, 5)$.

384. Find a 3×3 matrix C with eigenvalues $\lambda_1 = 1$, $\lambda_2 = 2$, $\lambda_3 = 3$ and corresponding eigenvectors $\mathbf{v}_1 = (1, 0, 1)$, $\mathbf{v}_2 = (1, 1, 2)$, $\mathbf{v}_3 = (1, 2, 4)$.

Diagonalizing Real Symmetric Matrices

Theorem 10.5: Let A be a real symmetric matrix.
 (i) Each root λ of the characteristic polynomial $\Delta(t)$ of A is real.
 (ii) If $\mathbf{u}_1$, $\mathbf{u}_2$, ..., $\mathbf{u}_r$ are eigenvectors of A belonging to distinct eigenvalues of A, then $\mathbf{u}_1$, $\mathbf{u}_2$, ..., $\mathbf{u}_r$ are orthogonal.

Theorem 10.6: Let A be a real symmetric matrix. Then there exists an orthogonal matrix P such that $D = P^{-1}AP$ is diagonal.

Remark 10.5: The orthogonal matrix P is obtained by normalizing a basis of orthogonal eigenvectors of A. The entries in the diagonal matrix D are the eigenvalues of A with their multiplicity.

Questions 385–387 refer to the symmetric matrix $A = \begin{bmatrix} 3 & 2 \\ 2 & 3 \end{bmatrix}$.

385. Find the characteristic polynomial $\Delta(t)$ and all eigenvalues of A.

386. Find a maximum set S of nonzero orthogonal eigenvectors of A.

387. Find an orthogonal matrix P and a diagonal matrix D such that $D = P^{-1}AP$.

Questions 388–390 refer to the symmetric matrix $B = \begin{bmatrix} 7 & 3 \\ 3 & -1 \end{bmatrix}$.

388. Find the characteristic polynomial $\Delta(t)$ and all eigenvalues of B.

389. Find a maximum set S of nonzero orthogonal eigenvectors of B.

390. Find an orthogonal matrix P and a diagonal matrix D such that $D = P^{-1}BP$.

Questions 391–393 refer to the symmetric matrix $C = \begin{bmatrix} 2 & 3 & 1 \\ 3 & 10 & 3 \\ 1 & 3 & 2 \end{bmatrix}$.

391. Find the characteristic polynomial $\Delta(t)$ and all eigenvalues of C.

392. Find a maximum set S of nonzero orthogonal eigenvectors of C.

393. Find an orthogonal matrix P and a diagonal matrix D such that $D = P^{-1}CP$.

Quadratic Forms

Definition: A polynomial q is called a *quadratic form* if q is a real polynomial in variables $x_1, x_2, \ldots, x_n$ such that every term has degree two—that is, if

$$q(x_1, x_2, \ldots, x_n) = a_{11}x_1^2 + a_{22}x_2^2 + \cdots + a_{nn}x_n^2 + \Sigma_{i<j}\, d_{ij}x_i x_j$$

or, equivalently, if $q(\mathbf{x}) = \mathbf{x}^T A \mathbf{x}$ where A is a symmetric matrix with $a_{ij} = d_{ij}/2$ and where $\mathbf{x} = [x_1, x_2, \ldots, x_n]^T$, the column vector of variables. If there are no cross-product terms $x_i y_j$ (i.e., all $d_{ij} = 0$), then q is said to be *diagonal*.

Questions 394–396: Find the quadratic form corresponding to each symmetric matrix.

394. $A = \begin{bmatrix} 5 & -3 \\ -3 & 8 \end{bmatrix}$

395. $B = \begin{bmatrix} 3 & & \\ & -4 & \\ & & 6 \end{bmatrix}$

396. $C = \begin{bmatrix} 2 & -5 & 1 \\ -5 & -6 & -7 \\ 1 & -7 & 9 \end{bmatrix}$

397. Find the symmetric matrix A corresponding to

$$q(x, y, z) = 3x^2 + 4xy - y^2 + 8xz - 6yz + z^2.$$

398. Find the symmetric matrix B corresponding to $q(x, y) = 4x^2 + 5xy - 7y^2$.

399. Find the symmetric matrix C corresponding to $q(x, y, z) = 4xy + 5z^2$.

400. Find the symmetric matrix D corresponding to $q(x, y, z) = x^2 - 2yz + xz$.

Questions 401–405 refer to the quadratic form $q(x, y) = 3x^2 + 2xy - y^2$ and to the linear substitution $L: x = s - 3t$ and $y = 2s + t$.

401. Find $q(s, t)$.

402. Find the matrix A corresponding to q, and rewrite q in matrix notation.

403. Find the matrix P corresponding to L, and rewrite L using matrix notation.

404. Find $q(s, t)$ using matrix notation.

405. Let $q(x, y) = 3x^2 - 12xy + 7y^2$. By "completing the square," find a linear substitution L expressing the variables x and y in terms of the variables s and t so that $q(s, t)$ is diagonal.

Minimum Polynomial

Definition: Let A be an n-square matrix, and let $J(A)$ be the collection of all polynomials for which A is a root. [By the Cayley-Hamilton Theorem, $J(A)$ is not empty.] Let $m(t)$ be the monic polynomial of minimal degree in $J(A)$. Then $m(t)$ is the *minimal polynomial* of A.

Theorem 10.7: The minimal polynomial $m(t)$ of a matrix A divides every polynomial $f(t)$ for which $f(A) = 0$. [In particular, $m(t)$ divides the characteristic polynomial $\Delta(t)$ of A.]

Theorem 10.8: The characteristic polynomial $\Delta(t)$ and the minimal polynomial $m(t)$ of A have the same irreducible factors.

Theorem 10.9: A scalar λ is an eigenvalue of A if and only if λ is a root of the minimal polynomial $m(t)$ of A.

Theorem 10.10: Let $M = \text{diag}(A_1, A_2, \ldots, A_n)$ be a block diagonal matrix. Then the minimal polynomial $m(t)$ of M is the least common multiple of the minimal polynomials of the diagonal blocks.

Remark 10.6: Let M be an n-square matrix with λ's on the diagonal, a's on the superdiagonal (where $a \neq 0$), and 0's elsewhere. Then $f(t) = (t - \lambda)^n$ is both the characteristic and the minimal polynomial of M.

Questions 406–409 refer to the matrices $A = \begin{bmatrix} 4 & -2 & 2 \\ 6 & -3 & 4 \\ 3 & -2 & 3 \end{bmatrix}$ and $B = \begin{bmatrix} 3 & -2 & 2 \\ 4 & -4 & 6 \\ 2 & -3 & 5 \end{bmatrix}$.

406. Find the characteristic polynomial $\Delta(t)$ of A.

407. Find the minimal polynomial $m(t)$ of A.

408. Find the characteristic polynomial $\Delta(t)$ of B.

409. Find the minimal polynomial $m(t)$ of B.

Questions 410–415 refer to the following matrices:

$$A = \begin{bmatrix} 2 & 5 & 0 & 0 & 0 \\ 0 & 2 & 0 & 0 & 0 \\ 0 & 0 & 4 & 2 & 0 \\ 0 & 0 & 3 & 5 & 0 \\ 0 & 0 & 0 & 0 & 7 \end{bmatrix}, B = \begin{bmatrix} 3 & 1 & 0 & 0 & 0 \\ 0 & 3 & 0 & 0 & 0 \\ 0 & 0 & 3 & 1 & 0 \\ 0 & 0 & 0 & 3 & 1 \\ 0 & 0 & 0 & 0 & 3 \end{bmatrix}, C = \begin{bmatrix} \lambda & & & & \\ & \lambda & & & \\ & & \lambda & & \\ & & & \lambda & \\ & & & & \lambda \end{bmatrix}.$$

410. Find the characteristic polynomial $\Delta(t)$ of A.

411. Find the minimal polynomial $m(t)$ of A.

412. Find the characteristic polynomial $\Delta(t)$ of B.

413. Find the minimal polynomial $m(t)$ of B.

414. Find the characteristic polynomial $\Delta(t)$ of C.

415. Find the minimal polynomial $m(t)$ of C.

Further Properties of Eigenvalues and Eigenvectors

416. Suppose λ is an eigenvalue of an invertible operator T. Show that λ^{-1} is an eigenvalue of T^{-1}.

417. Suppose $\mathbf{v}$ is a nonzero eigenvector of linear maps S and T. Show that $\mathbf{v}$ is an eigenvector of $S + T$.

418. Suppose $\mathbf{v}$ is an eigenvector of T. Show that, for any scalar k, $\mathbf{v}$ is an eigenvector of kT.

419. Suppose λ is an eigenvalue of T.
Prove (A) λ^2 is an eigenvalue of T^2. (B) λ is an eigenvalue of T^n.

420. Suppose λ is an eigenvalue of T. Show that $f(\lambda)$ is an eigenvalue of $f(T)$ for any polynomial $f(t)$.

CHAPTER **11**

Canonical Forms

Invariant Subspaces

Definition: Let $T: V \to V$ be linear. A subspace W of V is said to be *invariant* under T or *T-invariant* if T maps W into W—that is, if for every $\mathbf{w} \in W$ we have $T(\mathbf{w}) \in W$. In this case, T restricted to W induces a linear operator $\hat{T}: W \to W$ defined by $\hat{T}(\mathbf{w}) = T(\mathbf{w})$.

Remark 11.1: Clearly $\{0\}$ and V itself are invariant under any linear map $T: V \to V$.

Questions 421–424 refer to the linear operator $T: \mathbf{R}^2 \to \mathbf{R}^2$ which rotates each vector about the z-axis by an angle θ, as pictured in Figure 11.1; that is,

$$T(x, y, z) = (x \cos \theta,\ y \sin \theta,\ z).$$

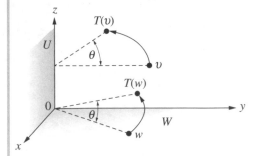

Figure 11.1

421. Let W be the xy-plane in $\mathbf{R}^2$. Is W invariant under T?

422. Let W' be the yz-plane in $\mathbf{R}^2$. Is W' invariant under T?

423. Let U be the z-axis in $\mathbf{R}^2$. Is U invariant under T?

424. Let U' be the x-axis in $\mathbf{R}^2$. Is U' invariant under T?

425. What, if any, is the relationship between eigenvectors of a linear operator T and invariant subspaces of T?

426. Show that the kernel of $T: V \rightarrow V$ is invariant under T.

427. Show that the image of $T: V \rightarrow V$ is invariant under T.

428. Find all invariant subspaces of $A = \begin{bmatrix} 4 & 2 \\ 3 & -1 \end{bmatrix}$ viewed as a linear operator on $\mathbf{R}^2$.

429. Find all invariant subspaces of $A = \begin{bmatrix} 2 & -4 \\ 5 & -2 \end{bmatrix}$ viewed as a linear operator on $\mathbf{R}^2$.

430. Find all invariant subspaces of $A = \begin{bmatrix} 2 & -4 \\ 5 & -2 \end{bmatrix}$ viewed as a linear operator on $\mathbf{C}^2$.

Direct-Sum Decompositions

Definition: A vector space V is termed the *direct sum* of its subspaces $W_1, \ldots, Wr$, written $V = W_1 \oplus W_2 \oplus \cdots \oplus W_r$, if every vector $\mathbf{v} \in V$ can be written uniquely in the form $\mathbf{v} = \mathbf{w}_1 + \mathbf{w}_2 + \cdots + \mathbf{w}_r$ with $\mathbf{w}_i \in W_i$. In such a case, the projection of V into its subspace W_i is the mapping $E: V \rightarrow V$ defined by $E(\mathbf{v}) = \mathbf{w}_i$.

Questions 431–434 refer to the following subspaces of $\mathbf{R}^3$:

$$U = xy\text{-plane}, \quad W = yz\text{-plane}, \quad Z = z\text{-axis}, \quad L = \{(k, k, k) \mid k \in \mathbf{R}\}.$$

431. Is $\mathbf{R}^3 = U \oplus W$?

432. Is $\mathbf{R}^3 = U \oplus Z$?

433. Is $\mathbf{R}^3 = U \oplus L$?

434. Is $\mathbf{R}^3 = W \oplus L$?

Nilpotent Operators and Matrices

Definition: A linear operator $T: V \rightarrow V$ is termed *nilpotent* if $T^n = 0$ for some positive integer n; we call k the *index of nilpotency* of T if $T^k = 0$ but $T^{k-1} \neq 0$.

Analogously, a square matrix A is termed *nilpotent* if $A^n = 0$ for some positive integer n, and of *index* k if $A^k = 0$ but $A^{k-1} \neq 0$.

Questions 435–438 refer to an n-square nilpotent matrix A of index k.

435. Find the minimal polynomial $m(t)$ of A.

436. Find the eigenvalues of A.

437. Show that $k \leq n$; that is, the index of A does not exceed the order of A.

438. Show that A is singular.

Questions 439–441 refer to the following matrices:

$$A = \begin{bmatrix} -2 & 1 & 1 \\ -3 & 1 & 2 \\ -2 & 1 & 1 \end{bmatrix}, \quad B = \begin{bmatrix} 1 & 3 & -2 \\ 1 & 3 & -2 \\ 1 & 3 & -2 \end{bmatrix}, \quad C = \begin{bmatrix} 1 & -3 & 2 \\ 1 & -3 & 2 \\ 1 & -3 & 2 \end{bmatrix}.$$

439. Is A nilpotent? If so, what is its index?

440. Is B nilpotent? If so, what is its index?

441. Is C nilpotent? If so, what is its index?

Definition: A *basic nilpotent block* N of index k is the k-square matrix with 1's on the superdiagonal and 0's elsewhere.

Remark 11.2: The basic nilpotent blocks of order 1, 2, 3, 4 follow:

$$[0], \quad \begin{bmatrix} 0 & 1 \\ 0 & 0 \end{bmatrix}, \quad \begin{bmatrix} 0 & 1 & 0 \\ 0 & 0 & 1 \\ 0 & 0 & 0 \end{bmatrix}, \quad \begin{bmatrix} 0 & 1 & 0 & 0 \\ 0 & 0 & 1 & 0 \\ 0 & 0 & 0 & 1 \\ 0 & 0 & 0 & 0 \end{bmatrix}.$$

Theorem 11.1: Every nilpotent matrix A is similar to a unique nilpotent block diagonal matrix $M = \text{diag}\,(N_1, N_2, \ldots, N_m)$ where each diagonal entry N_i is a basic nilpotent block.

442. Find the canonical nilpotent form of matrix A in Problem 439.

443. Find the canonical nilpotent form of matrix B in Problem 440.

444. Find the canonical nilpotent form of matrix C in Problem 441.

Jordan Canonical Form

Definition: A *Jordan block J* of order k belonging to an eigenvalue λ is the k-square matrix with λ's on the diagonal, with 1's on the superdiagonal, and with 0's elsewhere.

Remark 11.3: The basic Jordan blocks of order 1, 2, 3, 4 belonging to eigenvalue $\lambda = 7$ follow:

$$[7], \quad \begin{bmatrix} 7 & 1 \\ 0 & 7 \end{bmatrix}, \quad \begin{bmatrix} 7 & 1 & 0 \\ 0 & 7 & 1 \\ 0 & 0 & 7 \end{bmatrix}, \quad \begin{bmatrix} 7 & 1 & 0 & 0 \\ 0 & 7 & 1 & 0 \\ 0 & 0 & 7 & 1 \\ 0 & 0 & 0 & 7 \end{bmatrix}.$$

445. Find the characteristic polynomial $\Delta(t)$ and minimal polynomial $m(t)$ of the above Jordan block A of order 4 belonging to the eigenvalue $\lambda = 7$. What are the eigenvalues of A?

446. Find a basis for the eigenspace of the eigenvalue $\lambda = 7$ for the matrix A in Problem 445.

Definition: A matrix M is a *Jordan matrix* if M is a block diagonal matrix $M = \text{diag}(J_1, J_2, \ldots, J_r)$ where each diagonal entry J_i is a basic Jordan block.

Questions 447–449 refer to the Jordan matrix:

$$M = \text{diag}\left(\begin{bmatrix} -3 & 1 & 0 \\ 0 & -3 & 1 \\ 0 & 0 & -3 \end{bmatrix}, \begin{bmatrix} 5 & 1 \\ 0 & 5 \end{bmatrix}, \begin{bmatrix} 5 & 1 \\ 0 & 5 \end{bmatrix} \right).$$

447. Find the characteristic polynomial $\Delta(t)$ and eigenvalues of M.

448. Find the minimal polynomial $m(t)$ of M.

449. Find a maximum set S of linearly independent eigenvectors of M.

Questions 450–458 refer to the Jordan matrices:

$$A = \text{diag}\left(\begin{bmatrix} 4 & 1 & 0 \\ 0 & 4 & 1 \\ 0 & 0 & 4 \end{bmatrix}, \begin{bmatrix} 4 & 1 \\ 0 & 4 \end{bmatrix}, \begin{bmatrix} 2 & 1 \\ 0 & 2 \end{bmatrix}, [2] \right),$$

$$B = \text{diag}\left(\begin{bmatrix} 4 & 1 \\ 0 & 4 \end{bmatrix}, \begin{bmatrix} 4 & 1 \\ 0 & 4 \end{bmatrix}, [4], \begin{bmatrix} 2 & 1 & 0 \\ 0 & 2 & 1 \\ 0 & 0 & 2 \end{bmatrix} \right).$$

450. Find the characteristic polynomial $\Delta(t)$ and eigenvalues of A.

451. Find the minimal polynomial $m(t)$ of A.

452. Find the dimension d_1 of the eigenspace E_1 of $\lambda_1 = 4$ in A. Also, find a basis of E_1.

453. Find the dimension d_2 of the eigenspace E_2 of $\lambda_2 = 2$ in A. Also, find a basis of E_2.

454. Find the characteristic polynomial $\Delta(t)$ and eigenvalues of B.

455. Decide whether A and B are equivalent Jordan matrices.

456. Find the minimal polynomial $m(t)$ of B.

457. Find the dimension d_1 of the eigenspace E_1 of $\lambda_1 = 4$ in B. Also, find a basis of E_1.

458. Find the dimension d_2 of the eigenspace E_2 of $\lambda_2 = 2$ in B. Also, find a basis of E_2.

459. Find all Jordan canonical forms of a matrix A with the following characteristic polynomial and minimal polynomial:

$$\Delta(t) = (t{-}2)^4 \, (t{-}3)^3 \text{ and } m(t) = (t{-}2)^2 \, (t{-}3)^2.$$

460. Find all Jordan canonical forms of a matrix A with the following characteristic polynomial and minimal polynomial:

$$\Delta(t) = (t{-}7)^5 \text{ and } m(t) = (t{-}7)^2.$$

CHAPTER 12

Linear Operators on Inner Product Spaces

This chapter investigates the space $A(V)$ of linear operators T on an inner product space V. Thus the base field is either the real numbers **R** or the complex numbers **C**.

Adjoint Operators

Definition: Let $T: V \rightarrow V$ be linear. The *adjoint* of T is a linear map T^* on V such that, for every $\mathbf{u}, \mathbf{v} \in V$, $\langle T(\mathbf{u}), \mathbf{v} \rangle = \langle \mathbf{u}, T^*(\mathbf{v}) \rangle$.

461. Let A be a real matrix. Show that the transpose A^{T} of A is the adjoint of A.

462. Let B be a complex matrix. Show that the conjugate transpose B^* of B is the adjoint of B.

Questions 463–465: Find the adjoint of each matrix:

463. $A = \begin{bmatrix} 2+3i & 5-4i \\ 6-9i & 2+7i \end{bmatrix}$

464. $B = \begin{bmatrix} 3-7i & 7i & 8-i \\ 18 & 6+i & 7-9i \\ 8+i & 7+9i & 6+3i \end{bmatrix}$

465. $C = \begin{bmatrix} 1 & 2 & 3 \\ 4 & 5 & 6 \\ 7 & 7 & 7 \end{bmatrix}$

Theorem 12.1: Let S and T be linear operators on V and let $k \in K$. Then

(i) $(S + T)^* = S^* + T^*$
(ii) $(kT)^* = \bar{k}T^*$
(iii) $(ST)^* = T^*S^*$
(iv) $(T^*)^* = T$

466. Prove Theorem 12.1(i).

467. Prove Theorem 12.1(ii).

468. Prove Theorem 12.1(iii).

469. Prove Theorem 12.1(iv).

470. Let T be a linear operator on V, and let W be a T-invariant subspace of V. Show that the orthogonal complement $W^{\perp}$ of W is invariant under T^*.

471. Use the definition of the adjoint to show that:

(A) $I^* = I$
(B) $0^* = 0$

472. Suppose T is invertible. Show that $(T^{-1})^* = (T^*)^{-1}$.

Table 12.1

Class of Complex Numbers	Behavior under Conjugation	Class of Operators in $A(V)$	Behavior under the Adjoint Map

Real axis	$\bar{z} = z$	Self-adjoint operators
		Also called:
		  Symmetric (real case)
		  Hermitian (complex case)
			$T^* = T$

Unit circle ($\|z\| = 1$)	$\bar{z} = 1/z$	Orthogonal operators (real case) Unitary operators (complex case)	$T^* = T^{-1}$

Imaginary axis	$\bar{z} = -z$	Skew-adjoint operators
		Also called:
		  Skew-symmetric (real case)
		  Skew-Hermitian (complex case)
			$T^* = -T$

Positive half axis $(0, \infty)$	$z = \bar{w}\, w,$ $w \neq 0$	Positive definite operators	$T = S^*S$ with S nonsingular

Remark 12.1: Table 12.1 shows the analogy between classes of complex numbers under conjugation $z \rightarrow \bar{z}$ and classes of linear operators under the adjoint mapping $T \rightarrow T^*$.

Self-Adjoint Operators, Symmetric Operators

Definition: An operator T on V is *self-adjoint* if $T = T^*$. The terms *symmetric* and *Hermitian* are also used for self-adjoint operators on V when the base fields are **R** and **C**, respectively.

Theorem 12.2: Suppose T is self-adjoint. Then any eigenvalue λ of T is real.

Theorem 12.3: Suppose T is self-adjoint. Then eigenvectors of T belonging to distinct eigenvalues are orthogonal.

Theorem 12.4: Let A be a real symmetric matrix. Then there exists an orthogonal matrix P such that $D = P^{-1}AP$ is diagonal.

473. Prove Theorem 12.2.

474. Prove Theorem 12.3.

475. Let $A = \begin{bmatrix} 5 & 4 \\ 4 & -1 \end{bmatrix}$. Find a maximum set S of linearly independent eigenvectors of A.

476. For the matrix A in Problem 475, find an orthogonal matrix P such that $P^{-1}AP$ is diagonal.

Problems 477–479 refer to diagonalizing the symmetric matrix $C = \begin{bmatrix} 2 & 2 & 1 \\ 2 & 5 & 2 \\ 1 & 2 & 2 \end{bmatrix}$.

477. Find the characteristic polynomial $\Delta(t)$ of C and all its eigenvalues.

478. Find a maximum set S of nonzero orthogonal eigenvectors of C.

479. Find an orthogonal matrix P and diagonal matrix D such that $D = P^{-1}CP$.

480. Consider the quadratic form $q(x, y, z) = 2x^2 + 4xy + 5y^2 + 2xz + 4yz + 2z^2$. Find an orthogonal change of coordinates that diagonalizes q.

481. Suppose T is self-adjoint and $T^2 = 0$. Show that $T = 0$.

482. Show that, for any operator T, the following are self-adjoint:
(A) T^*T and T^*T
(B) $T + T^*$

Orthogonal and Unitary Operators

Definition: Let U be an invertible operator on V such that $U* = U^{-1}$ or, equivalently, $UU^* = U^*U = I$. Then U is said to be *orthogonal* or *unitary* according to whether the base field is **R** or **C**.

Theorem 12.5: The following conditions on an operator U are equivalent:
  (i) U is unitary (orthogonal); that is, $U^* = U^{-1}$.
 (ii) U preserves inner products; that is, $\forall \mathbf{u}, \mathbf{v} \in V, \langle U(\mathbf{u}), U(\mathbf{w}) \rangle = \langle \mathbf{u}, \mathbf{w} \rangle$.
(iii) U preserves lengths; that is, $\forall \mathbf{v} \in V, \|U(\mathbf{v})\| = \|\mathbf{v}\|$.

Theorem 12.6: Let T be an orthogonal operator on V. Then there exists an orthogonal basis B of V such that the matrix representation of T in B has the following block diagonal form:

$$[T]_B = \text{diag}\left(I_p, -I_q, \begin{bmatrix} \cos\theta_1 & -\sin\theta_1 \\ \sin\theta_1 & \cos\theta_1 \end{bmatrix}, \ldots, \begin{bmatrix} \cos\theta_r & -\sin\theta_r \\ \sin\theta_r & \cos\theta_r \end{bmatrix} \right).$$

483. Suppose $T: \mathbf{R}^3 \to \mathbf{R}^3$ rotates each vector **v** about the *z*-axis by a fixed angle θ, as pictured in Figure 12.1(a). Show that T is orthogonal.

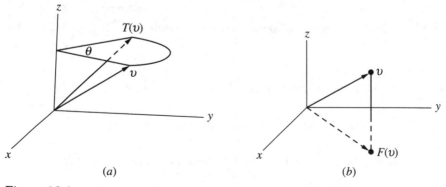

(a)   *(b)*

Figure 12.1

484. Suppose $F: \mathbf{R}^3 \to \mathbf{R}^3$ reflects each vector $\mathbf{v}$ through the xy-plane; that is, suppose $F(x, y, z) = (x, y, -z)$, as pictured in Figure 12.1(b). Show that F is orthogonal.

485. Suppose T is unitary (orthogonal) and λ is an eigenvalue of T. Show that $|\lambda| = 1$.

486. Show that every 2×2 orthogonal matrix A for which $\det(A) = 1$ has the following form for some real number θ: $\begin{bmatrix} \cos\theta & -\sin\theta \\ \sin\theta & \cos\theta \end{bmatrix}$.

Positive and Positive Definite Operators

Definition: A linear operator on V is said to be *positive* (or *semi-definite*) if $P = S^*S$ for some operator S, and is said to be *positive definite* if S is nonsingular.

487. Show that a positive (positive definite) operator P is self-adjoint.

Theorem 12.7: The following are equivalent for an operator P on V.
   (i) P is positive; that is, $P = S^*S$.
  (ii) $P = T^2$ where T is self-adjoint.
 (iii) P is self-adjoint and $\langle P(\mathbf{u}), \mathbf{u} \rangle \geq 0$ for every $\mathbf{u}$ in V.

Theorem 12.8: The following are equivalent for an operator P on V.
   (i) P is positive definite; that is, $P = S^*S$ for some nonsingular S.
  (ii) $P = T^2$ where T is nonsingular and self-adjoint.
 (iii) P is self-adjoint and $\langle P(\mathbf{u}), \mathbf{u} \rangle$ is positive for every $\mathbf{u} \neq \mathbf{0}$ in V.

Theorem 12.9: A complex matrix $A = \begin{bmatrix} a & b \\ c & d \end{bmatrix}$ is positive (positive definite) if and only if A is self-adjoint, and a, d, and $|A| = ad - bc$ are nonnegative (positive).

488. Determine whether each matrix is positive definite or positive:

(A) $A = \begin{bmatrix} 1 & 1 \\ 1 & 1 \end{bmatrix}$

(B) $B = \begin{bmatrix} 3 & i \\ -i & 3 \end{bmatrix}$

(C) $C = \begin{bmatrix} 1 & 1 \\ 0 & 1 \end{bmatrix}$

489. Determine whether each matrix is positive definite or positive:

(A) $D = \begin{bmatrix} 2 & 1 \\ 1 & 2 \end{bmatrix}$

(B) $E = \begin{bmatrix} 1 & i \\ -i & 1 \end{bmatrix}$

(C) $F = \begin{bmatrix} 1 & -2i \\ 2i & 1 \end{bmatrix}$

490. Let λ be an eigenvalue of T. Show that: (a) if T is positive, then λ is real and nonnegative; (b) if T is positive definite, then λ is real and positive.

Questions 491–494 refer to the matrix $A = \begin{bmatrix} 5 & 1 \\ 1 & 5 \end{bmatrix}$.

491. Show that A is positive definite.

492. Find an orthogonal matrix Q such that $Q^{-1} AQ$ is diagonal.

493. Find the square root S of $B = Q^{T} AQ = \begin{bmatrix} 6 & 0 \\ 0 & 4 \end{bmatrix}$.

494. Find the positive square root T of A.

Normal Operators

Definition: A linear operator T on V is *normal* if T commutes with its adjoint; that is, $TT^* = T^*T$.

Remark 12.2: Thus a complex matrix A is normal if $AA^* = A^*A$ and a real matrix A is normal if $AA^{T} = A^{T}A$.

Questions 495–497 refer to matrices $A = \begin{bmatrix} 1 & 1 \\ i & 3+2i \end{bmatrix}$, $B = \begin{bmatrix} 1 & i \\ 0 & 1 \end{bmatrix}$, $C = \begin{bmatrix} 1 & i \\ 1 & 2+i \end{bmatrix}$.

495. Determine whether A is normal.

496. Determine whether B is normal.

497. Determine whether C is normal.

498. Suppose T is normal. Show that $T - \lambda I$ is normal.

499. Suppose $\mathbf{u}$ and $\mathbf{v}$ are eigenvectors of a normal operator T belonging to distinct eigenvalues. Say, $T(\mathbf{v}) = \lambda_1 \mathbf{v}$ and $T(\mathbf{w}) = \lambda_2 \mathbf{w}$ and $\lambda_1 \neq \lambda_2$. Show that $\langle \mathbf{v}, \mathbf{w} \rangle = 0$; i.e., $\mathbf{v}$ and $\mathbf{w}$ are orthogonal.

Definition: A linear operator T on an inner product space V is said to be *diagonalizable* if there exist operators $E_1, ..., E_r$ and scalars $\lambda_1, ..., \lambda_r$ such that
- (i) $T = \lambda_1 E_1 + \lambda_2 E_2 + ... + \lambda_r E_r$
- (ii) $E_1 + E_2 + ... + E_r = I$
- (iii) $E_1^2 = E_1, ..., E_r^2 = E_r$
- (iv) $E_i E_j = 0$ for $i \neq j$

Theorem 12.10 (Spectral Theorem): Let T be a normal (symmetric) operator on a complex (real) finite-dimensional inner product space V. Then T is diagonalizable by the above definition.

500. Show that a diagonal matrix $A = \mathrm{diag}(2, 3, 3, 5)$ is diagonalizable by the above definition.

ANSWERS

Chapter 1: Vectors

1. **(A)** The vectors **u** and **v** are equal if and only if corresponding components are equal.
   (B) Only $\mathbf{u}_2$ and $\mathbf{u}_4$ are componentwise equal.

2. **(A)** Add corresponding components:

   $$(3, -4, 5, -6) + (1, 1, -2, 4) = (3 + 1, -4 + 1, 5 - 2, -6 + 4) = (4, -3, 3, -2).$$

   (B) The sum is not defined, since the vectors have different numbers of components.
   (C) Multiply each component by the scalar: $-3(4, -5, -6) = (-12, 15, 18)$.
   (D) Take the negative of each component: $(6, -7, 8)$.

3. **(A)** Add corresponding components:

   $$\begin{bmatrix} 7 \\ -4 \\ 2 \end{bmatrix} + \begin{bmatrix} -3 \\ -1 \\ 5 \end{bmatrix} = \begin{bmatrix} 7 - 3 \\ -4 - 1 \\ 2 + 5 \end{bmatrix} = \begin{bmatrix} 4 \\ -5 \\ 7 \end{bmatrix}.$$

   (B) Multiply each component by the scalar:

   $$5 \begin{bmatrix} -2 \\ 3 \\ 4 \end{bmatrix} = \begin{bmatrix} -10 \\ 15 \\ 20 \end{bmatrix}.$$

4. Perform first the scalar multiplication and then the vector addition.
   (A) $3\mathbf{u} - 4\mathbf{v} = 3(2, -7, 1) - 4(-3, 0, 4) = (6, -21, 3) + (12, 0, -16) = (18, -21, -13)$
   (B) $2\mathbf{u} + 3\mathbf{v} - 5\mathbf{w} = 2(2, -7, 1) + 3(-3, 0, 4) - 5(0, 5, -8)$

   $$= (4, -14, 2) + (-9, 0, 12) + (0, -25, 40)$$

   $$= (4 - 9 + 0, -14 + 0 - 25, 2 + 12 + 40) = (-5, -39, 54)$$

5. We want to find scalars x and y such that $\mathbf{w} = x\mathbf{u} + y\mathbf{v}$; i.e.,

   $$(1, 9) = x(1, 2) + y(3, -1) = (x, 2x) + (3y, -y) = (x + 3y, 2x - y).$$

Equality of corresponding components gives the two equations

   $$x + 3y = 1, \quad 2x - y = 9.$$

To solve the system of equations, multiply the first equation by -2 and then add the result to the second equation to obtain $-7y = 7$, or $y = -1$. Then substitute $y = -1$ in the first equation to obtain $x - 3 = 1$, or $x = 4$. Accordingly, $\mathbf{w} = 4\mathbf{u} - \mathbf{v}$.

6. Proceed as in Problem 5, this time using column representations:

$$\begin{bmatrix} 2 \\ -3 \\ 4 \end{bmatrix} = x\begin{bmatrix} 1 \\ 1 \\ 1 \end{bmatrix} + y\begin{bmatrix} 1 \\ 1 \\ 0 \end{bmatrix} + z\begin{bmatrix} 1 \\ 0 \\ 0 \end{bmatrix} = \begin{bmatrix} x \\ x \\ x \end{bmatrix} + \begin{bmatrix} y \\ y \\ 0 \end{bmatrix} + \begin{bmatrix} z \\ 0 \\ 0 \end{bmatrix} = \begin{bmatrix} x+y+z \\ x+y \\ x \end{bmatrix}$$

Now set corresponding components equal to each other:

$$x + y + z = 2, \qquad x + y = -3, \qquad x = 4.$$

To solve the system of equations, substitute $x = 4$ in the second equation to obtain $4 + y = -3$, or $y = -7$. Then substitute in the first equation to find $z = 5$. Thus, $\mathbf{v} = 4\mathbf{u}_1 - 7\mathbf{u}_2 + 5\mathbf{u}_3$.

7. (A) $S_n = f(1) + f(2) + \cdots + f(n-1) + f(n)$. Note, for $n \geq 2$, $S_n = S_{n-1} + f(n)$.

(B) $\sum_{k=n_1}^{n_2} f(k) = f(n_1) + f(n_1 + 1) + f(n_1 + 2) + \cdots + f(n_2)$. For $n_2 < n_1$, the summation

is usually defined to be zero.

8. (A) $\sum_{k=1}^{4} k^3 = 1^3 + 2^3 + 3^3 + 4^3 = 1 + 8 + 27 + 64 = 100.$

(B) $\sum_{j=2}^{5} j^2 = 2^2 + 3^2 + 4^2 + 5^2 = 4 + 9 + 16 + 25 = 54.$

9. Multiply corresponding components and add:
(A) $\mathbf{u} \cdot \mathbf{v} = (2)(8) + (-3)(2) + (6)(-3) = -8.$
(B) $\mathbf{u} \cdot \mathbf{v} = (3)(4) + (-5)(1) + (2)(-2) + (1)(5) = 8.$
(C) $\mathbf{u} \cdot \mathbf{v} = (1)(6) + (-2)(7) + (3)(1) + (-4)(-2) = 3.$

10. (A) First find $\|\mathbf{u}\|^2 = \mathbf{u} \cdot \mathbf{u}$ by squaring the components of $\mathbf{u}$ and adding:

$$\|\mathbf{u}\|^2 = 3^2 + (-12)^2 + (-4)^2 = 9 + 144 + 16 = 169.$$

Then, $\|\mathbf{u}\| = \sqrt{169} = 13$.
(B) Square each component of v and then add to obtain $\|\mathbf{v}\|^2 = \mathbf{v} \cdot \mathbf{v}$:

$$\|\mathbf{v}\|^2 = 2^2 + (-3)^2 + 8^2 + (-5)^2 = 4 + 9 + 64 + 25 = 102.$$

Then $\|\mathbf{v}\| = \sqrt{102}$.
(C) $\|\mathbf{w}\|^2 = (-3)^2 + 1^2 + (-2)^2 + 4^2 + (-5)^2 = 9 + 1 + 4 + 16 + 25 = 55$; hence $\|\mathbf{w}\| = \sqrt{55}$.

11. $\|\mathbf{u}\|^2 = 1^2 + k^2 + (-2)^2 + 5^2 = k^2 + 30$. Now solve $k^2 + 30 = 39$ and obtain $k = 3, -3$.

12. The vector $\hat{\mathbf{v}}$ is a unit vector, since

$$\hat{\mathbf{v}} \cdot \hat{\mathbf{v}} = \left(\frac{\mathbf{v}}{\|\mathbf{v}\|}\right)\left(\frac{\mathbf{v}}{\|\mathbf{v}\|}\right) = \frac{1}{\|\mathbf{v}\|^2}(\mathbf{v} \cdot \mathbf{v}) = \frac{1}{\|\mathbf{v}\|^2}\|\mathbf{v}\|^2 = 1.$$

Moreover, $\hat{\mathbf{v}}$ is in the same direction as $\mathbf{v}$, since $\hat{\mathbf{v}}$ is a positive scalar multiple of $\mathbf{v}$.

13. (A) First find $\|\mathbf{v}\|^2 = \mathbf{v} \cdot \mathbf{v} = 12^2 + (-3)^2 + (-4)^2 = 144 + 9 + 16 = 169$. Then divide each component of $\mathbf{v}$ by $\|\mathbf{v}\| = \sqrt{169} = 13$ to obtain

$$\hat{\mathbf{v}} = \frac{\mathbf{v}}{\|\mathbf{v}\|} = \left(\frac{12}{13}, \frac{-3}{13}, \frac{-4}{13}\right).$$

(B) First find $\|\mathbf{w}\|^2 = \mathbf{w} \cdot \mathbf{w} = 4^2 + (-2)^2 + (-3)^2 + 8^2 = 16 + 4 + 9 + 64 = 93$. Divide each component of $\mathbf{w}$ by $\|\mathbf{w}\| = \sqrt{93}$ to obtain

$$\hat{\mathbf{w}} = \frac{\mathbf{w}}{|\mathbf{w}|} = \left(\frac{4}{\sqrt{93}}, \frac{-2}{\sqrt{93}}, \frac{-3}{\sqrt{93}}, \frac{8}{\sqrt{93}}\right).$$

14. $\mathbf{v}$ and any positive multiple of $\mathbf{v}$ will have the same normalized form. Hence, first multiply $\mathbf{v}$ by 12 to "clear" fractions: $12\mathbf{v} = (6, 8, -3)$. Then $\|12\mathbf{v}\|^2 = 36 + 64 + 9 = 109$. Accordingly, the required unit vector is

$$\hat{\mathbf{v}} = 12\mathbf{v} = \frac{12\mathbf{v}}{\|12\mathbf{v}\|} = \left(\frac{6}{\sqrt{109}}, \frac{8}{\sqrt{109}}, \frac{-3}{\sqrt{109}}\right).$$

15. By the Cauchy–Schwarz inequality (Theorem 1.1) and other properties of the inner product,

$$\|\mathbf{u} + \mathbf{v}\|^2 = (\mathbf{u} + \mathbf{v}) \cdot (\mathbf{u} + \mathbf{v}) = \mathbf{u} \cdot \mathbf{u} + 2(\mathbf{u} \cdot \mathbf{v}) + \mathbf{v} \cdot \mathbf{v}$$
$$\leq \|\mathbf{u}\|^2 + 2\|\mathbf{u}\|\|\mathbf{v}\| + \|\mathbf{v}\|^2 = (\|\mathbf{u}\| + \|\mathbf{v}\|)^2.$$

Taking the square root of both sides yields the desired inequality.

16. In each case use the formula $d(\mathbf{u}, \mathbf{v}) = \|\mathbf{u} - \mathbf{v}\| = \sqrt{(u_1 - v_1)^2 + \cdots + (u_n - v_n)^2}$.

(A) $d(\mathbf{u}, \mathbf{v}) = \sqrt{(1-6)^2 + (7+5)^2} = \sqrt{25 + 144} = \sqrt{169} = 13.$

(B) $d(\mathbf{u}, \mathbf{v}) = \sqrt{(3-6)^2 + (-5-2)^2 + (4+1)^2} = \sqrt{9 + 49 + 25} = \sqrt{83}.$

17. $(d(\mathbf{u}, \mathbf{v}))^2 = \|\mathbf{u} - \mathbf{v}\|^2 = (2-3)^2 + (k+1)^2 + (1-6)^2 + (-4+3)^2 = k^2 + 2k + 28$. Now solve $k^2 + 2k + 28 = 6^2$ to obtain $k = 2, -4$.

18. By Cauchy–Schwarz inequality, $|\mathbf{u} \cdot \mathbf{v}| \leq \|\mathbf{u}\|\|\mathbf{v}\|$. Hence $-1 \leq \theta \leq 1$, which uniquely defines a real angle $0 \leq \theta \leq \pi$.

19. (A) First find

$$\mathbf{u} \cdot \mathbf{v} = 3 + 10 - 21 = -8, \|\mathbf{u}\|^2 = 1 + 4 + 9 = 14, \|\mathbf{v}\|^2 = 9 + 25 + 49 = 83.$$

Then, using the formula in the *cross product* definition,

$$\cos\theta = \frac{\mathbf{u} \cdot \mathbf{v}}{\|\mathbf{u}\|\|\mathbf{v}\|} = -\frac{8}{\sqrt{14}\sqrt{83}}.$$

(B) $\mathbf{u} \cdot \mathbf{v} = 8 - 18 - 1 + 20 = 9, \|\mathbf{u}\|^2 = 16 + 9 + 1 + 25 = 51, \|\mathbf{v}\|^2 = 4 + 36 + 1 + 16 = 57.$

Then $\cos\theta = \dfrac{9}{\sqrt{51}\sqrt{57}}.$

20. (A) First find $\mathbf{u} \cdot \mathbf{v} = 2 - 10 + 12 = 4$ and $\|\mathbf{v}\|^2 = 4 + 25 + 16 = 45$. Then, by the *cross*

product definition, $\text{proj}(\mathbf{u}, \mathbf{v}) = \dfrac{\mathbf{u} \cdot \mathbf{v}}{\|\mathbf{v}\|^2} \mathbf{v} = \dfrac{4}{45}(2, 5, 4) = \left(\dfrac{8}{45}, \dfrac{20}{45}, \dfrac{16}{45}\right) = \left(\dfrac{8}{45}, \dfrac{4}{9}, \dfrac{16}{45}\right)$.

(B) First find $\mathbf{u} \cdot \mathbf{v} = 12 - 18 - 4 + 5 = -5$ and $\|\mathbf{v}\|^2 = 9 + 36 + 16 + 1 = 62$. Then

$$\text{proj}(\mathbf{u} \cdot \mathbf{v}) = \dfrac{\mathbf{u} \cdot \mathbf{v}}{\|\mathbf{v}\|^2} \mathbf{v} = -\dfrac{5}{62}(3, 6, -4, 1) = \left(\dfrac{-15}{62}, \dfrac{-30}{62}, \dfrac{20}{62}, \dfrac{-5}{62}\right) = \left(\dfrac{-15}{62}, \dfrac{-15}{31}, \dfrac{10}{31}, \dfrac{-5}{62}\right).$$

[Observe that when $\mathbf{u} \cdot \mathbf{v} < 0$, $\text{proj}(\mathbf{u}, \mathbf{v})$ is in the opposite direction to $\mathbf{v}$.]

21. Find the dot product of each pair of vectors:

$$\mathbf{u} \cdot \mathbf{v} = 15 - 16 + 1 = 0, \ \mathbf{v} \cdot \mathbf{w} = 3 + 8 + 3 = 14, \ \mathbf{u} \cdot \mathbf{w} = 5 - 8 + 3 = 0.$$

Hence $\mathbf{u}$ and $\mathbf{v}$ are orthogonal, $\mathbf{u}$ and $\mathbf{w}$ are orthogonal, but $\mathbf{v}$ and $\mathbf{w}$ are not orthogonal.

22. Solve $\mathbf{u} \cdot \mathbf{v} = (1)(2) + (k)(-5) + (-3)(4) = 2 - 5k - 12 = 0$ for k, obtaining $k = -2$.

23. (A) $(\mathbf{u} + \mathbf{v}) \cdot \mathbf{w} = \mathbf{u} \cdot \mathbf{w} + \mathbf{v} \cdot \mathbf{w} = 0 + 0 = 0$

(B) $(k\mathbf{u}) \cdot \mathbf{w} = k(\mathbf{u} \cdot \mathbf{w}) = k0 = 0$

24. Take the dot product of $\mathbf{w}$ with $\mathbf{u}_1$ to obtain

$$\mathbf{w} \cdot \mathbf{u}_1 = (x\mathbf{u}_1 + y\mathbf{u}_2 + z\mathbf{u}_3) \cdot \mathbf{u}_1 = x(\mathbf{u}_1 \cdot \mathbf{u}_1) + y(\mathbf{u}_2 \cdot \mathbf{u}_1) + z(\mathbf{u}_3 \cdot \mathbf{u}_1)$$
$$= x(\mathbf{u}_1 \cdot \mathbf{u}_1) + y(0) + z(0) = x(\mathbf{u}_1 \cdot \mathbf{u}_1) \equiv x \|\mathbf{u}_1\|^2$$

whence $x = (\mathbf{w} \cdot \mathbf{u}_1)/\|\mathbf{u}_1\|^2$. Similarly, take the dot product of $\mathbf{w}$ with $\mathbf{u}_2$ to obtain $y = (\mathbf{w} \cdot \mathbf{u}_2)/\|\mathbf{u}_2\|^2$, and take the dot product of $\mathbf{w}$ with $\mathbf{u}_3$ to obtain $z = (\mathbf{w} \cdot \mathbf{u}_3)/\|\mathbf{u}_3\|^2$.

25. Compute the dot product of each pair of vectors:

$$\mathbf{u}_1 \cdot \mathbf{u}_2 = 1 - 4 + 3 = 0, \ \mathbf{u}_1 \cdot \mathbf{u}_3 = -8 - 4 + 12 = 0, \ \mathbf{u}_2 \cdot \mathbf{u}_3 = -8 + 4 + 4 = 0.$$

Thus the vectors are orthogonal to each other.

26. Since $\mathbf{u}_1, \mathbf{u}_2, \mathbf{u}_3$ are orthogonal to each other, use the formulas of Problem 24.

$\mathbf{w} \cdot \mathbf{u}_1 = 13 + 8 + 21 = 42$	$\|\mathbf{u}_1\|^2 = 1 + 4 + 9 = 14$	$x = 42/14 = 3$
$\mathbf{w} \cdot \mathbf{u}_2 = 13 - 8 + 7 = 12$	$\|\mathbf{u}_2\|^2 = 1 + 4 + 1 = 6$	$y = 12/6 = 2$
$\mathbf{w} \cdot \mathbf{u}_3 = -104 - 8 + 28 = -84$	$\|\mathbf{u}_3\|^2 = 64 + 4 + 16 = 84$	$z = -84/84 = -1$

27. (A) To find $\mathbf{u} + \mathbf{v}$, add the corresponding components, yielding

$$\mathbf{u} + \mathbf{v} = 7\mathbf{i} + 2\mathbf{j} + 5\mathbf{k}.$$

(B) To find $3\mathbf{u} - 2\mathbf{v}$, first multiply the vectors by the scalars and then add:

$$3\mathbf{u} - 2\mathbf{v} = (9\mathbf{i} + 15\mathbf{j} - 6\mathbf{k}) + (-8\mathbf{i} + 6\mathbf{j} - 14\mathbf{k}) = \mathbf{i} + 21\mathbf{j} - 20\mathbf{k}.$$

(C) To find $\mathbf{u} \cdot \mathbf{v}$, multiply the corresponding components and then add:

$$\mathbf{u} \cdot \mathbf{v} = 12 - 15 - 14 = -17.$$

(D) To find $\|\mathbf{u}\|$, square each component and then add to get $\|\mathbf{u}\|^2$:

$$\|\mathbf{u}\|^2 = 9 + 25 + 4 = 38 \text{ and hence } \|\mathbf{u}\| = \sqrt{38}.$$

28. (A) $\begin{vmatrix} 3 & 4 \\ 5 & 9 \end{vmatrix} = (3)(9) - (4)(5) = 7$

(B) $\begin{vmatrix} 2 & -1 \\ 4 & 3 \end{vmatrix} = 6 + 4 = 10$

(C) $\begin{vmatrix} 4 & 5 \\ 3 & -2 \end{vmatrix} = -8 - 15 = -23$

29. [*Hint:* $-\begin{vmatrix} a & b \\ c & d \end{vmatrix} = -(ad - bc) = bc - ad$, the product bc of the nondiagonal elements minus the product ad of the diagonal elements. This is called taking the determinant backward.]

(A) $-\begin{vmatrix} 3 & 6 \\ 4 & 2 \end{vmatrix} = (6)(4) - (3)(2) = 18$

(B) $-\begin{vmatrix} 7 & -5 \\ 3 & 2 \end{vmatrix} = -15 - 14 = -29$

(C) $-\begin{vmatrix} 4 & -1 \\ 8 & -3 \end{vmatrix} = -8 + 12 = 4$

(D) $-\begin{vmatrix} -4 & -3 \\ 2 & -2 \end{vmatrix} = -6 - 8 = -14$

30. (A) $\mathbf{u} \times \mathbf{v} = \left(\begin{vmatrix} \boxed{1} & 2 & 3 \\ \boxed{4} & 5 & 6 \end{vmatrix}, -\begin{vmatrix} 1 & \boxed{2} & 3 \\ 4 & \boxed{5} & 6 \end{vmatrix}, \begin{vmatrix} 1 & 2 & \boxed{3} \\ 4 & 5 & \boxed{6} \end{vmatrix} \right) = (12 - 15, 12 - 6, 5 - 8)$

$= (-3, 6, -3)$

(B) $\mathbf{u} \times \mathbf{v} = \left(\begin{vmatrix} \boxed{7} & 3 & 1 \\ \boxed{1} & 1 & 1 \end{vmatrix}, -\begin{vmatrix} 7 & \boxed{3} & 1 \\ 1 & \boxed{1} & 1 \end{vmatrix}, \begin{vmatrix} 7 & 3 & \boxed{1} \\ 1 & 1 & \boxed{1} \end{vmatrix} \right) = (3 - 1, 1 - 7, 7 - 3) = (2, -6, 4)$

31. $\mathbf{u} \cdot (\mathbf{u} \times \mathbf{v}) = a_1(a_2 b_3 - a_3 b_2) + a_2(a_3 b_1 - a_1 b_3) + a_3(a_1 b_2 - a_2 b_1)$
$= a_1 a_2 b_3 - a_1 a_3 b_2 + a_2 a_3 b_1 - a_1 a_2 b_3 + a_1 a_3 b_2 - a_2 a_3 b_1 = 0.$

Thus, $\mathbf{u} \times \mathbf{v}$ is orthogonal to $\mathbf{u}$. Similarly, $\mathbf{u} \times \mathbf{v}$ is orthogonal to $\mathbf{v}$.

32. In view of Problem 31, first find $\mathbf{u} \times \mathbf{w}$. The array

$$\begin{bmatrix} 1 & 3 & 4 \\ 2 & -6 & -5 \end{bmatrix} \text{ gives } \mathbf{v} \times \mathbf{w} = (-15 + 24, 8 + 5, -6 - 6) = (9, 13, -12).$$

Now normalize $\mathbf{v} \times \mathbf{w}$ to get $\mathbf{u} = (9/\sqrt{394}, 13/\sqrt{394}, -12/\sqrt{394})$.

33. Use the ordinary rules of algebra together with $i^2 = -1$ to obtain a result in the standard form $a + bi$.

    (A) $z + w = (2 + 3i) + (2 + 3i) + (5 - 2i) = 7 + i$
    (B) $z - w = (2 + 3i) - (5 - 2i) = 2 + 3i - 5 + 2i = -3 + 5i$
    (C) $zw = (2 + 3i)(5 - 2i) = 10 - 4i + 15i - 6i^2 = 16 + 11i$

34. **(A)** $i^{39} = i^{4 \cdot 9 + 3} = (i^4)^9 i^3 = 1^9 i^3 = i^3 = -i$
    (B) $i^{174} = i^2 = -1$
    (C) $i^{252} = i^0 = 1$
    (D) $i^{317} = i^1 = i$

35. **(A)** $6 - 4i$
    (B) $7 + 5i$
    (C) $4 - i$
    (D) $-3 + i$

36. **(A)** $4 + 9 = 13$
    (B) $16 + 25 = 41$
    (C) $36 + 4 = 40$

37. Use $|z|^2 \ z\bar{z} = a^2 + b^2$ where $z = a + bi$.

    (A) $z\bar{z} = 3^2 + 4^2 = 25, \qquad |z| = \sqrt{25} = 5$
    (B) $z\bar{z} = 5^2 + (-2)^2 = 29, \qquad |z| = \sqrt{29}$
    (C) $z\bar{z} = (-7)^2 + 1^2 = 50, \qquad |z| = \sqrt{50} = 5\sqrt{2}$
    (D) $z\bar{z} = (-1)^2 + (-4)^2 = 17, \qquad |z| = \sqrt{17}$

38. To simplify a fraction z/w of the complex numbers, multiply both numerator and denominator by $\bar{w}$, the conjugate of the denominator.

    (A) $\dfrac{1}{3 - 4i} = \dfrac{(3 + 4i)}{(3 - 4i)(3 + 4i)} = \dfrac{3 + 4i}{25} = \dfrac{3}{25} + \dfrac{4}{25}i$

    (B) $\dfrac{2 - 7i}{5 + 3i} = \dfrac{(2 - 7i)(5 - 3i)}{(5 + 3i)(5 - 3i)} = \dfrac{-11 - 41i}{34} = -\dfrac{11}{34} - \dfrac{41}{34}i$

39. **(A)** $\mathbf{u} + \mathbf{v} = (7 - 2i + 1 + i, 2 + 5i - 3 - 6i) = (8 - i, -1 - i)$
    (B) $(3 - i)\mathbf{v} = (3 + 3i - i - i^2, -9 - 18i + 3i + 6i^2) = (4 + 2i, -15 - 15i)$
    (C) $(1 + i)\mathbf{u} + (2 - i)\mathbf{v} = (9 + 5i, -3 + 6i) + (3 + i, -12 - 3i) = (12 + 6i, -15 + 3i)$

40. (A) $\mathbf{u}\cdot\mathbf{v}=(2+3i)(\overline{3-2i})+(4-i)(\overline{5})+(2i)(\overline{4-6i})$

$=(2+3i)(3+2i)+(4-i)(5)+(2i)(4+6i)=13i+20-5i-12+8i=8+16i.$

(B) $\mathbf{u}\cdot\mathbf{u}=(2+3i)(\overline{2+3i})+(4-i)(\overline{4-i})+(2i)(\overline{2i})$

$=(2+3i)(2-3i)+(4-i)(4+i)+(2i)(-2i)=13+17+4=34.$

So $\|\mathbf{u}\|=\sqrt{\mathbf{u}\cdot\mathbf{u}}=\sqrt{34}.$

(C) $\mathbf{v}\cdot\mathbf{v}=(3-2i)(3+2i)+(5)(5)+(4-6i)(4+6i)=13+25+52=90.$

So $\|\mathbf{v}\|=\sqrt{90}=3\sqrt{10}.$

Chapter 2: Matrices

41. The *rows* of A are the horizontal lines of numbers; there are two of them: $[1\quad 2\quad 3]$ and $[4\quad 5\quad 6]$. The *columns* of A are the vertical lines of numbers; there are three of them:

$$\begin{bmatrix}1\\4\end{bmatrix}\quad\begin{bmatrix}2\\5\end{bmatrix}\quad\begin{bmatrix}3\\6\end{bmatrix}.$$

The *size* of A is 2×3 (read "2 by 3"), the number of rows by the number of columns.

42. A and B are equal if and only if they have the same size and corresponding entries are equal.

43. Equate corresponding entries:

$$x+y=3,\qquad x-y=1,\qquad 2z+w=5,\qquad z-w=4.$$

The solution of the system of equations is $x=2,\ y=1,\ z=3,\ w=-1.$

44. Perform first the scalar multiplications and then a matrix addition:

$$2A-3B=\begin{bmatrix}2 & -4 & 6\\8 & 10 & -12\end{bmatrix}+\begin{bmatrix}-9 & 0 & -6\\21 & -3 & -24\end{bmatrix}=\begin{bmatrix}-7 & -4 & 0\\29 & 7 & -36\end{bmatrix}.$$

(Note that we multiply B by -3 and then add, rather than multiplying B by 3 and subtracting. This helps us avoid errors.)

45. Multiply corresponding entries and add.

(A) $(8,-4,5)\begin{bmatrix}3\\2\\-1\end{bmatrix}=(8)(3)+(-4)(2)+(5)(-1)=24-8-5=11$

(B) $(6,-1,7,5)\begin{bmatrix}4\\-9\\-3\\2\end{bmatrix}=24+9-21+10=22$

46. Since A is 2×2 and B is 2×3, the product AB is defined as a 2×3 matrix. To obtain the entries in the first row of AB, multiply the first row [1 3] of A by the columns $\begin{bmatrix} 2 \\ 3 \end{bmatrix}, \begin{bmatrix} 0 \\ -2 \end{bmatrix}$, and $\begin{bmatrix} -4 \\ 6 \end{bmatrix}$ of B, respectively:

$$\begin{bmatrix} \boxed{\begin{matrix} 1 & 3 \end{matrix}} \\ 2 & -1 \end{bmatrix} \begin{bmatrix} \boxed{2} & \boxed{0} & \boxed{-4} \\ 3 & -2 & 6 \end{bmatrix}$$

$$[2+9 \quad 0-6 \quad -4+18]$$
$$= [11 \quad -6 \quad 14)].$$

To obtain the entries in the second row of AB, multiply the second row $[2, -1]$ of A by the columns of B, respectively:

$$\begin{bmatrix} 1 & 3 \\ \boxed{\begin{matrix} 2 & -1 \end{matrix}} \end{bmatrix} \begin{bmatrix} \boxed{2} & \boxed{0} & \boxed{-4} \\ 3 & -2 & 6 \end{bmatrix} =$$

$$[4-3, 0+2, -8-6] = [1, 2, -14]$$

Thus

$$AB = \begin{bmatrix} 11 & -6 & 14 \\ 1 & 2 & -14 \end{bmatrix}.$$

47. Since A is 2×3 and B is 3×4, the product is defined as a 2×4 matrix. Multiply the rows of A by the columns of B to obtain

$$AB = \begin{bmatrix} 4+3-4 & -2+9-1 & 0-15+2 & 12+3-2 \\ 8-2+20 & -4-6+5 & 0+10-10 & 24-2+10 \end{bmatrix} = \begin{bmatrix} 3 & 6 & -13 & 13 \\ 26 & -5 & 0 & 32 \end{bmatrix}.$$

48. Let $A = \begin{bmatrix} 1 & 6 \\ -3 & 5 \end{bmatrix}$ and $B = \begin{bmatrix} 4 & 0 \\ 2 & -1 \end{bmatrix}$. Then

$$AB = \begin{bmatrix} 1 & 6 \\ -3 & 5 \end{bmatrix} \begin{bmatrix} 4 & 0 \\ 2 & -1 \end{bmatrix} = \begin{bmatrix} 4+12 & 0-6 \\ -12+10 & 0-5 \end{bmatrix} = \begin{bmatrix} 16 & -6 \\ -2 & -5 \end{bmatrix}$$

$$BA = \begin{bmatrix} 4 & 0 \\ 2 & -1 \end{bmatrix} \begin{bmatrix} 1 & 6 \\ -3 & 5 \end{bmatrix} = \begin{bmatrix} 4+0 & 24+0 \\ 2+3 & 12-5 \end{bmatrix} = \begin{bmatrix} 4 & 24 \\ 5 & 7 \end{bmatrix}.$$

Matrix multiplication does not obey the commutative law.

49. The first and second rows of A become the first and second columns of A^T:

$$A^T = \begin{bmatrix} 1 & 4 \\ 2 & -5 \\ 3 & -6 \end{bmatrix}.$$

Equivalently, the first, second, and third columns of A become the first, second, and third rows of A^T.

50. Rewrite the rows of A as columns to obtain A^{T}, and then rewrite the rows of A^{T} as columns to obtain $(A^{\mathrm{T}})^{\mathrm{T}}$:

$$A^{\mathrm{T}} = \begin{bmatrix} 1 & 6 \\ 3 & -7 \\ 5 & -8 \end{bmatrix}, \quad (A^{\mathrm{T}})^{\mathrm{T}} = \begin{bmatrix} 1 & 3 & 5 \\ 6 & -7 & -8 \end{bmatrix}.$$

Observe that $(A^{\mathrm{T}})^{\mathrm{T}} = A$.

51. (A) $AB = \begin{bmatrix} 5-12 & 0+14 \\ 15+24 & 0-28 \end{bmatrix} = \begin{bmatrix} -7 & 14 \\ 39 & -28 \end{bmatrix}$, so $(AB)^{\mathrm{T}} = \begin{bmatrix} -7 & 39 \\ 14 & -28 \end{bmatrix}$.

(B) We have

$$A^{\mathrm{T}} = \begin{bmatrix} 1 & 3 \\ 2 & -4 \end{bmatrix} \quad B^{T} = \begin{bmatrix} 5 & -6 \\ 0 & 7 \end{bmatrix}.$$

Then $\qquad A^{\mathrm{T}}B^{\mathrm{T}} = \begin{bmatrix} 5+0 & -6+21 \\ 10+0 & -12-28 \end{bmatrix} = \begin{bmatrix} 5 & 15 \\ 10 & -40 \end{bmatrix}.$

Note that $(AB)^{\mathrm{T}} \neq A^{\mathrm{T}}B^{\mathrm{T}}$.

52. (A) Interchanging the same two rows twice, we obtain the original matrix; that is, this operation is its own inverse.

(B) Multiplying the ith row by k and then by k^{-1}, or by k^{-1} and then by k, we obtain the original matrix. In other words, the operations $\mathbf{R}_i \to k\mathbf{R}_i$ and $\mathbf{R}_i \to k^{-1}\mathbf{R}_i$ are inverses.

(C) Applying the operation $\mathbf{R}_i \to k\mathbf{R}_j + \mathbf{R}_i$, and then the operation $\mathbf{R}_i \to -k\mathbf{R}_j + \mathbf{R}_i$, or applying the operation $\mathbf{R}_i \to -k\mathbf{R}_j + \mathbf{R}_i$ and then the operation $\mathbf{R}_i \to k\mathbf{R}_j + \mathbf{R}_i$, we obtain the original matrix. In other words, the operations $\mathbf{R}_i \to k\mathbf{R}_j + \mathbf{R}_i$ and $\mathbf{R}_i \to -k\mathbf{R}_j + \mathbf{R}_i$ are inverses.

53. E is equivalent to E_2 (with parameter k) followed by E_3 (with parameter k').

54. (A) $\begin{bmatrix} 1 & 2 & 3 & 4 \\ 3 & -4 & 5 & -6 \\ 5 & 6 & 7 & 8 \end{bmatrix}$

(B) $\begin{bmatrix} 3 & 6 & 9 & 12 \\ 5 & 6 & 7 & 8 \\ 3 & -4 & 5 & -6 \end{bmatrix}$

(C) $\begin{bmatrix} 1 & 2 & 3 & 4 \\ 5 & 6 & 7 & 8 \\ 0 & -10 & -4 & -18 \end{bmatrix}$

55. The new scalar in the kj-position of A is
   (A) $(-a_{kj}/a_{ij})a_{ij} + a_{kj} = 0$,
   (B) $(-a_{kj})a_{ij} + (a_{ij})a_{kj} = 0$.

56. First write down $\mathbf{R}_1$, since it will not change:

$$[2 \quad 1 \quad -3 \quad 4].$$

To obtain a 0 in $\mathbf{R}_2$ below the pivot, apply the operation $\mathbf{R}_2 \rightarrow -3\mathbf{R}_1 + 2\mathbf{R}_2$, that is, calculate

$$\begin{bmatrix} 2 & 1 & -3 & 4 \\ -6+6 & -3+8 & 9+2 & -12-4 \end{bmatrix} = \begin{bmatrix} 2 & 1 & -3 & 4 \\ 0 & 5 & 11 & -16 \end{bmatrix}.$$

To obtain a 0 in R_3 below the pivot, apply the operation $\mathbf{R}_3 \rightarrow -5\mathbf{R}_1 + 2\mathbf{R}_3$:

$$\begin{bmatrix} 2 & 1 & -3 & 4 \\ 0 & 5 & 11 & -16 \\ -10+10 & -5-4 & 15+6 & -20+0 \end{bmatrix} = \begin{bmatrix} 2 & 1 & -3 & 4 \\ 0 & 5 & 11 & -16 \\ 0 & -9 & 21 & -20 \end{bmatrix}.$$

This is the required matrix.

57. The leading nonzero entries are the first nonzero entries in the rows of the matrix.

(A) $\begin{bmatrix} 0 & \boxed{1} & -3 & 4 & 6 \\ \boxed{4} & 0 & 2 & 5 & -3 \\ 0 & 0 & \boxed{7} & -2 & 8 \end{bmatrix}$

(B) $\begin{bmatrix} 0 & 0 & 0 & 0 & 0 \\ \boxed{1} & 2 & 3 & 4 & 5 \\ 0 & 0 & \boxed{5} & -4 & 7 \end{bmatrix}$

(C) $\begin{bmatrix} 0 & \boxed{2} & 2 & 2 & 2 \\ 0 & \boxed{3} & 1 & 0 & 0 \\ 0 & 0 & 0 & 0 & 0 \end{bmatrix}$

58. None of the three matrices is in echelon form. (In the third matrix, the 3 is not to the right of the 2.)

59. Use $a_{11} = 1$ as a pivot to obtain 0s below a_{11}; that is, apply the row operations $\mathbf{R}_2 \rightarrow -2\mathbf{R}_1 + \mathbf{R}_2$ and $\mathbf{R}_3 \rightarrow -3\mathbf{R}_1 + \mathbf{R}_3$ to obtain the matrix

$$\begin{bmatrix} 1 & 2 & -3 & 0 \\ 0 & 0 & 4 & 2 \\ 0 & 0 & 5 & 3 \end{bmatrix}.$$

Now use $a_{23} = 4$ as a pivot to obtain a 0 below a_{23}; that is, apply the row operation $R_3 \to -5R_2 + 4R_3$ to obtain the matrix

$$\begin{bmatrix} 1 & 2 & -3 & 0 \\ 0 & 0 & 4 & 2 \\ 0 & 0 & 0 & 2 \end{bmatrix}$$

which is in echelon form.

60. First interchange R_1 and R_2 to obtain a nonzero pivot in the first row; then apply $R_3 \to -R_1 + R_3$; and finally apply $R_3 \to -2R_2 + R_3$:

$$A \sim \begin{bmatrix} 2 & 1 & -4 & 2 \\ 0 & 1 & 3 & -2 \\ 2 & 3 & 2 & -1 \end{bmatrix} \sim \begin{bmatrix} 2 & 1 & -4 & 3 \\ 0 & 1 & 3 & -2 \\ 0 & 2 & 6 & -4 \end{bmatrix} \sim \begin{bmatrix} 2 & 1 & -4 & 3 \\ 0 & 1 & 3 & -2 \\ 0 & 0 & 0 & 0 \end{bmatrix}$$

The matrix is now in echelon form.

61. (A) The matrix is not in row canonical form since leading nonzero entries are not 1.
 (B) The leading nonzero entry in the second row is not the only nonzero entry in its column; thus the matrix is not in row canonical form.
 (C) This matrix is in row canonical form.

62. Multiply R_3 by ¼ so that the leading nonzero entry, a_{35}, equals 1. Produce 0s above a_{35} by applying the operations $R_2 \to -5R_3 + R_2$ and $R_1 \to -6R_3 + R_1$:

$$A \sim \begin{bmatrix} 2 & 3 & 4 & 5 & 6 \\ 0 & 0 & 3 & 2 & 5 \\ 0 & 0 & 0 & 0 & 1 \end{bmatrix} \sim \begin{bmatrix} 2 & 3 & 4 & 5 & 0 \\ 0 & 0 & 3 & 2 & 0 \\ 0 & 0 & 0 & 0 & 1 \end{bmatrix}.$$

Multiply R_2 by ⅓ so that the leading nonzero entry, a_{23}, equals 1. Produce a 0 above a_{23} with the operation $R_1 \to -4R_2 + R_1$:

$$A \sim \begin{bmatrix} 2 & 3 & 4 & 5 & 0 \\ 0 & 0 & 1 & \frac{2}{3} & 0 \\ 0 & 0 & 0 & 0 & 1 \end{bmatrix} \sim \begin{bmatrix} 2 & 3 & 0 & \frac{7}{3} & 0 \\ 0 & 0 & 1 & \frac{2}{3} & 0 \\ 0 & 0 & 0 & 0 & 1 \end{bmatrix}.$$

Finally, multiply R_1 by ½ to obtain the row canonical form

$$A \sim \begin{bmatrix} 1 & \frac{3}{2} & 0 & \frac{7}{6} & 0 \\ 0 & 0 & 1 & \frac{2}{3} & 0 \\ 0 & 0 & 0 & 0 & 1 \end{bmatrix}.$$

63. First, reduce B to an echelon form by applying $R_2 \to -2R_1 + R_2$ and $R_3 \to -3R_1 + R_3$ and then $R_3 \to -R_2 + R_3$:

$$B \sim \begin{bmatrix} 2 & 2 & -1 & 6 & 4 \\ 0 & 0 & 3 & -2 & 5 \\ 0 & 0 & 3 & 2 & 7 \end{bmatrix} \sim \begin{bmatrix} 2 & 2 & -1 & 6 & 4 \\ 0 & 0 & 3 & -2 & 5 \\ 0 & 0 & 0 & 4 & 2 \end{bmatrix}.$$

Now apply Step 2 of the Gaussian algorithm. Multiply R_3 by ¼, making the pivot $b_{34} = 1$, and then apply $R_2 \to 2R_3 + R_2$ and $R_1 \to -6R_3 + R_1$:

$$B \sim \begin{bmatrix} 2 & 2 & -1 & 6 & 4 \\ 0 & 0 & 3 & -2 & 5 \\ 0 & 0 & 0 & 1 & \frac{1}{2} \end{bmatrix} \sim \begin{bmatrix} 2 & 2 & -1 & 0 & 1 \\ 0 & 0 & 3 & 0 & 6 \\ 0 & 0 & 0 & 1 & \frac{1}{2} \end{bmatrix}.$$

Now multiply R_2 by ⅓, making the pivot $b_{23} = 1$, and apply $R_1 \to R_2 + R_1$:

$$B \sim \begin{bmatrix} 2 & 2 & -1 & 0 & 1 \\ 0 & 0 & 1 & 0 & 2 \\ 0 & 0 & 0 & 1 & \frac{1}{2} \end{bmatrix} \sim \begin{bmatrix} 2 & 2 & 0 & 0 & 3 \\ 0 & 0 & 1 & 0 & 2 \\ 0 & 0 & 0 & 1 & \frac{1}{2} \end{bmatrix}.$$

Finally, multiply R_1 by ½ to obtain the row canonical form

$$B \sim \begin{pmatrix} 1 & 1 & 0 & 0 & \frac{3}{2} \\ 0 & 0 & 1 & 0 & 2 \\ 0 & 0 & 0 & 1 & \frac{1}{2} \end{pmatrix}.$$

64. An arbitrary matrix A may be row equivalent to many echelon matrices. On the other hand, regardless of the algorithm that is used, a matrix A is row equivalent to a unique matrix in row canonical form. (The term "canonical" usually denotes uniqueness.)

65. Multiplying R_n by $1/a_{nn}$ and using the new $a_{nn} = 1$ as a pivot, we convert the last column of A into the unit vector $[0, 0, \ldots, 1]^T$. Each succeeding back-substitution yields a new unit vector with 1 on the diagonal, and the end result is the identity matrix. Thus A has the $n \times n$ identity matrix I as its row canonical form.

66. These are $\begin{bmatrix} 1 & 0 \\ 0 & 1 \end{bmatrix}$, $\begin{bmatrix} 1 & k \\ 0 & 0 \end{bmatrix}$, $\begin{bmatrix} 0 & 0 \\ 0 & 0 \end{bmatrix}$, where k is an arbitrary scalar.

67. (A) The block matrix has two rows, $\begin{bmatrix} 1 & -2 & | & 0 & 1 & | & 3 \\ 2 & 3 & | & 5 & 7 & | & -2 \end{bmatrix}$ and $[3 \ \ 1 \ | \ 4 \ \ 5 \ | \ 9]$, and three columns,

$$\begin{bmatrix} 1 & -2 \\ 2 & 3 \\ \hline 3 & 1 \end{bmatrix} \quad \begin{bmatrix} 0 & 1 \\ 5 & 7 \\ \hline 4 & 5 \end{bmatrix} \quad \begin{bmatrix} 3 \\ -2 \\ \hline 9 \end{bmatrix}.$$

Therefore, its size is 2×3. (There are four *block sizes*: 2×2, 2×1, 1×2, and 1×1.)
   (B) 3×2

68. (A) The sum $A + B$ may be obtained by adding corresponding blocks:

$$A + B = \begin{bmatrix} A_{11} + B_{11} & A_{12} + B_{12} & \cdots & A_{1n} + B_{1n} \\ A_{21} + B_{21} & A_{22} + B_{22} & \cdots & A_{2n} + B_{2n} \\ \cdots\cdots\cdots\cdots\cdots\cdots\cdots\cdots\cdots\cdots\cdots \\ A_{m1} + B_{m1} & A_{m2} + B_{m2} & \cdots & A_{mn} + B_{mn} \end{bmatrix}.$$

The justification is that adding the corresponding blocks adds the corresponding elements of A and B.

(B) $kA = \begin{bmatrix} kA_{11} & kA_{12} & \cdots & kA_{1n} \\ kA_{21} & kA_{22} & \cdots & kA_{2n} \\ \cdots\cdots\cdots\cdots\cdots\cdots\cdots\cdots \\ kA_{m1} & kA_{m2} & \cdots & kA_{mn} \end{bmatrix}$

because multiplying each block by k effects the multiplication of each element of A by k.

69. The product UV may be obtained by multiplying the corresponding block matrices; that is,

$$UV = \begin{bmatrix} W_{11} & W_{12} & \cdots & W_{1n} \\ W_{21} & W_{22} & \cdots & W_{2n} \\ \cdots\cdots\cdots\cdots\cdots\cdots\cdots \\ W_{m1} & W_{m2} & \cdots & W_{mn} \end{bmatrix} \qquad \text{where} \qquad W_{ij} = U_{i1}V_{1j} + U_{i2}V_{2j} + \cdots + U_{ip}V_{pj}.$$

To convince yourself of the validity of block multiplication, consider the following computation of the $(1, 1)$-element of UV:

$$w_{11} = \overbrace{u_{11}v_{11} + u_{12}v_{21} + u_{13}v_{31} + \cdots}^{(1,1)\text{-element of } U_{11}V_{11}} + \overbrace{u_{1r}v_{r1} + v_{1,\,r+1} + u_{1,\,r+1}v_{r+1,\,1} + \cdots}^{(1,1)\text{-element of } U_{12}V_{21}} + \cdots\cdots\cdots.$$

Thus, the partitioning of U and V merely partitions the sums defining the elements of UV.

70. Here $A = \begin{bmatrix} E & F \\ 0_{1\times2} & G \end{bmatrix}$ and $B = \begin{bmatrix} R & S \\ 0_{1\times3} & T \end{bmatrix}$ where E, F, G, R, S, and T are the given blocks. Hence

$$AB = \begin{bmatrix} ER & ES + FT \\ 0_{1\times3} & GT \end{bmatrix} = \left(\begin{bmatrix} 9 & 12 & 15 \\ 19 & 26 & 33 \\ [\,0 & 0 & 0\,] \end{bmatrix} \begin{bmatrix} 3 \\ 7 \\ [2] \end{bmatrix} + \begin{bmatrix} 1 \\ 0 \\ \end{bmatrix} \right) = \begin{bmatrix} 9 & 12 & 15 & 4 \\ 19 & 26 & 33 & 7 \\ 0 & 0 & 0 & 2 \end{bmatrix}.$$

Chapter 3: Systems of Linear Equations

71. (A) When it has the standard form $a_1x_1 + a_2x_2 + \cdots + a_nx_n = b$.

The constant a_k is called the *coefficient* of x_k, and b is called the *constant* of the equation.

(B) When all coefficients equal 0.

72. As it stands, there are four unknowns: x, y, z, k. Because of the term ky, it is not a linear equation. However, assuming k is a constant, the equation is linear in the unknowns x, y, z.

73. In general, the n-tuple (vector in $\mathbf{R}^n$) $\mathbf{u} = (u_1, u_2, \ldots, u_n)$ is a solution of (or *solves or satisfies*) the equation $F(x_1, x_2, \ldots, x_n) = 0$ if $F(u_1, u_2, \ldots, u_n) = 0$. So we substitute in the given equation to obtain

$$3 + 2(2) - 3(1) \stackrel{?}{=} 4, \quad \text{or} \quad 3 + 4 - 3 \stackrel{?}{=} 4, \quad \text{or} \quad 4 \stackrel{?}{=} 4.$$

Yes, it is a solution.

74. (A) Rewrite in standard form by collecting terms and transposing:

$$2x + 3y - 3 = 2x + 3y, \text{ or } 2x + 3y - 2x - 3y = 3, \text{ or } 0x + 0y = 3.$$

The equation is degenerate with a nonzero constant; thus the equation has no solution.

(B) Rewrite in standard form by collecting terms and transposing:

$$y - x + 3 = y - x + 3, \quad \text{or} \quad y - x - y + x = 3 - 3, \quad \text{or} \quad 0x + 0y = 0.$$

The equation is degenerate with a zero constant; thus every vector $\mathbf{u} = (a, b)$ in $\mathbf{R}^2$ is a solution.

75. (A) Multiply by ¼ to obtain the unique solution $x = -12/4 = -3$.
    (B) Multiply by $1/5$ to obtain the unique solution $x = 0/5 = 0$.
    (C) Rewrite the equation in standard form $x - 5 = x + 3$, or $x - x = 3 + 8$, or $0x = 8$. The equation has no solution [Theorem 3.2(ii)].
    (D) Rewrite the equation in standard form $x + 1 = x + 1$, or $x - x = 1 + 1$, or $0x = 0$. Every scalar k is a solution [Theorem 3.2(iii)].

76. (A) Choose any value for either unknown, say $x = -2$. Substitute $x = -2$ in the equation to obtain $2(-2) + y = 4$, or $-4 + y = 4$, or $y = 8$. Thus $x = -2$, $y = 8$; in other words, the point $(-2, 8)$ in $\mathbf{R}^2$ is a solution. Now substitute $x = 3$ in the equation to obtain $2(3) + y = 4$, or $6 + y = 4$, or $y = -2$; hence $(3, -2)$ is a solution. Lastly, substitute $y = 0$ in the equation to obtain $2x + 0 = 4$, or $2x = 4$, or $x = 2$. Thus $(2, 0)$ is a solution.
    (B) Plot the three solutions $(-2, 8)$, $(3, -2)$, and $(2, 0)$ in the Cartesian plane $\mathbf{R}^2$, as pictured in Figure 3.1. Draw the straight line L determined by two of the solutions—say,

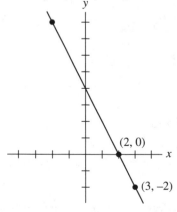

Figure 3.1

(–2, 8) and (2, 0)—and note that the third solution also lies on L. Indeed, L is the set of *all* solutions; that is to say, L is the graph of the given equation.

77. (A) The lines corresponding to the linear equations L_1 and L_2 intersect in a single point (Figure 3.2).
　　(B) The lines corresponding to the linear equations L_1 and L_2 are parallel (Figure 3.3).
　　(C) The lines corresponding to the linear equations L_1 and L_2 coincide (Figure 3.4).

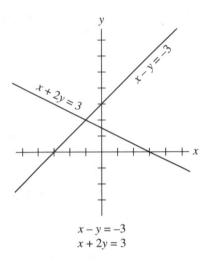

$$x - y = -3$$
$$x + 2y = 3$$

Figure 3.2

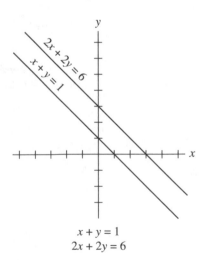

$$x + y = 1$$
$$2x + 2y = 6$$

Figure 3.3

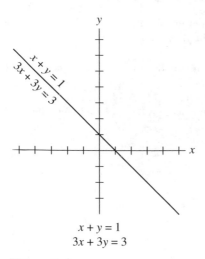

$$x + y = 1$$
$$3x + 3y = 3$$

Figure 3.4

78. To eliminate x, multiply $\mathbf{L}_1$ by 3 and multiply $\mathbf{L}_2$ by -2, and then add the resulting equations:

$$
\begin{array}{ll}
3\mathbf{L}_1: & 6x + 15y = 24 \\
-2\mathbf{L}_2: & -6x + 4y = 14 \\
\hline
\text{Addition:} & 19y = 38, \text{ or } y = 2
\end{array}
$$

Substitute $y = 2$ in one of the original equations, say $\mathbf{L}_1$, to obtain $2x + 5(2) = 8$, or $2x + 10 = 8$, or $2x = -2$, or $x = -1$. Hence $x = -1$ and $y = 2$, or the pair $(-1, 2)$, is the unique solution to the system.

79. By the *leading unknown* in a linear equation we mean the first unknown with a nonzero coefficient. Its position p is therefore the smallest integral value of j for which $a_j \neq 0$. Here, x_3 is the leading unknown, since $a_1 = 0$, $a_2 = 0$, but $a_3 \neq 0$. Thus $p = 3$.

80. (A) Here y is the leading unknown, and x and z are free variables. (1) Set $x = 1$ and $y = 1$. Substitution yields $y = 1/3$; hence $\mathbf{u}_1 = (1, 1/3, 1)$ is a solution. (2) Set $x = 0$ and $y = 1$. Substitution yields $y = 1/3$; hence $\mathbf{u}_2 = (0, 1/3, 1)$ is a solution. (3) Set $x = 1$ and $y = 0$. Substitution yields $y = 5/3$; hence $\mathbf{u}_3 = (1, 5/3, 0)$ is a solution.

    (B) Set $x = a$ and $z = b$ (where a and b are parameters). Substitute to obtain $3y - 4b = 5$, or $3y = 5 + 4b$, or $y = (5 + 4b)/3$. Thus $\mathbf{u} = (a, (5 + 4b)/3, b)$ is the general solution.

81. Although each equation shows only two unknowns, the system has three unknowns, x, y, and z. (We assume that there is no unknown with only zero coefficients.)

82. (A) Substitute **v** in each equation to obtain

(1) $-8+2(4)-5(1)+4(2) \overset{?}{=} 3,$ or $-8+8-5+8 \overset{?}{=} 3,$ or $3 \overset{?}{=} 3;$

(2) $2(-8)+3(4)+1-2(2) \overset{?}{=} 1,$ or $-16+14+1-4 \overset{?}{=} 1,$ or $-5 \overset{?}{=} 3.$

No, the second equation is not satisfied.

    (B) Substitute **u** in each equation to obtain

(1) $-8+2(6)-5(1)+4(1) \overset{?}{=} 3,$ or $-8+12-5+4 \overset{?}{=} 3,$ or $3 \overset{?}{=} 3;$

(2) $2(-8)+3(6)+1-2(1) \overset{?}{=} 1$ or $-16+18+1-2 \overset{?}{=} 1,$ or $1 \overset{?}{=} 1.$

    Yes, **u** is a solution.

83. (A) L_1: $x - 2y + 3z = 5$
      L_3: $3x + 2y - 7z = 3$
      L_2: $2x + \ \ y - 4z = 1$

    (B) L_1: $x - 2y + \ 3z = 5$
     $3L_2$: $6x + 3y - 12z = 3$
      L_3: $3x + 2y - \ 7z = 3$

    (C) Multiply L_1 by -3 and add it to L_3.

$$
\begin{aligned}
-3L_1: \quad & -3x + 6y - 9z = -15 \\
L_3: \quad & \ \ 3x + 2y - 7z = \ \ \ 3 \\
\hline
\text{Addition:} \quad & \ \ \ \ \ \ \ \ \ \ \ 8y - 16z = -12
\end{aligned}
$$

This last equation replaces the third equation in the original system to yield

$$
\begin{aligned}
L_1: \quad & x - 2y + \ 3z = \ \ 5 \\
L_2: \quad & 2x + \ \ y - \ 4z = \ \ 1 \\
-3L_1 + L_3: \quad & \ \ \ \ \ \ \ \ \ \ 8y - 16z = -12.
\end{aligned}
$$

(Observe that the unknown x has been eliminated from the third equation.)

84. (A) The equation does not have a solution, so the system does not have any solution.
    (B) Every vector satisfies the equation; hence we can delete the equation from the system without changing its solution set.

85. The system is in triangular form; hence we solve by back-substitution.
    (1) The last equation gives $t = 4$.
    (2) Substituting in the third equation gives $7z - 4 = 3$, or $7z = 7$, or $z = 1$.

(3) Substituting $z = 1$ and $t = 4$ in the second equation gives $5y - 1 + 3(4) = 1$, or $5y - 1 + 12 = 1$, or $5y = -10$, or $y = -2$.

(4) Substituting $y = -2$, $z = 1$, $t = 4$ in the first equation gives $2x - 3(-2) + 5(1) - 2(4) = 9$, or $2x + 6 + 5 - 8 = 9$, or $2x = 6$, or $x = 3$. Thus, $x = 3$, $y = -2$, $z = 1$, $t = 4$ is the unique solution of the system.

86. Here y and t are the free variables.

87. The system is in echelon form. The leading unknowns are x and z; hence the free variables are y and t. Accordingly, assign any values to y and t, and then solve by back-substitution for x and z to obtain a solution. For example:

(1) Let $y = 1$ and $t = 1$. Substitute $t = 1$ in the last equation to obtain $z - 4 = 2$, or $z = 6$. Substitute $y = 1$, $z = 6$, $t = 1$ in the first equation to obtain $x + 4(1) - 3(6) + 2(1) = 5$, or $x + 4 - 18 + 2 = 5$, or $x = 22$. Thus $\mathbf{u}_1 = (22, 1, 6, 1)$ is a particular solution.

(2) Let $y = 1$, $t = 0$. Substitute $t = 0$ in the last equation to get $z = 2$. Substitute $y = 1$, $z = 2$, $t = 0$ in the first equation to get $x = 6$. Thus $\mathbf{u}_2 = (6, 1, 2, 0)$ is a particular solution.

(3) Let $y = 0$, $t = 1$. Substitute $t = 1$ in the last equation to get $z = 6$. Substitute $y = 0$, $z = 6$, $t = 1$ in the first equation to get $x = 21$. Thus $\mathbf{u}_3 = (21, 0, 6, 1)$ is a particular solution.

88. **(A)** Use back-substitution to solve for the leading unknowns x and z in terms of the free variables y and t. The last equation gives $z = 2 + 4t$. Substitute in the first equation to obtain $x + 4y - 3(2 + 4t) + 2t = 5$, or $x + 4y - 6 - 12t + 2t = 5$, or $x = 11 - 4y + 10t$. Accordingly,

$$x = 11 - 4y + 10t$$
$$z = 2 + 4t$$

is the free-variable form of the general solution.

(B) In (A), set $y = a$ and $t = b$ to obtain the parametric form

$$x = 11 - 4a + 10b, \, y = a, \, z = 2 + 4b, \, t = b.$$

The solution vector is $\mathbf{u} = (11 - 4a + 10b, \, a, \, 2 + 4b, \, b)$.

89. If $\mathbf{L} = k\mathbf{L}'$ for some other equation $\mathbf{L}'$ in the system, the operation $\mathbf{L} \rightarrow -k\mathbf{L}' + \mathbf{L}$ replaces $\mathbf{L}$ by $0x_1 + 0x_2 + \cdots + 0x_n = 0$, which will be deleted under Step 3(a). In other words, in both cases $\mathbf{L}$ is deleted from the system.

90. **(A)** Reduce to echelon form. To eliminate x from the second and third equations, apply the operations $\mathbf{L}_2 \rightarrow -3\mathbf{L}_1 + 2\mathbf{L}_2$ and $\mathbf{L}_3 \rightarrow -5\mathbf{L}_1 + 2\mathbf{L}_3$.

$-3\mathbf{L}_1$:	$-6x - 3y + 6z = -30$	
$2\mathbf{L}_2$:	$6x + 4y + 4z = 2$	
$-3\mathbf{L}_1 + 2\mathbf{L}_2$:	$y + 10z = -28$	

$-5\mathbf{L}_1$:	$-10x - 5y + 10z = -50$	
$2\mathbf{L}_3$:	$10x + 8y + 6z = 8$	
$-5\mathbf{L}_1 + 2\mathbf{L}_3$:	$3y + 16z = -42$	

This yields the following system, from which y is eliminated from the third equation by the operation $L_3 \rightarrow -3L_2 + L_3$.

$$\left.\begin{array}{r} 2x + y - 2z = 10 \\ y + 10z = -28 \\ 3y + 16z = -42 \end{array}\right\} \rightarrow \left\{\begin{array}{r} 2x + y - 2z = 10 \\ y + 10z = -28 \\ -14z = 42. \end{array}\right.$$

The system is now in triangular form, and hence has the unique solution (back-substitution) $\mathbf{u} = (1, 2, -3)$.

(B) Reduce to echelon form. To eliminate x from the second and third equations, apply the operations $L_2 \rightarrow -2L_1 + L_2$ and $L_3 \rightarrow -3L_1 + L_3$ to obtain (using Problem 89)

$$\left.\begin{array}{r} x + 2y - 3z = 1 \\ y - 2z = 2 \\ 2y - 4z = 4 \end{array}\right\} \rightarrow \left\{\begin{array}{r} x + 2y - 3z = 1 \\ y - 2z = 2. \end{array}\right.$$

The system is now in echelon form, with free variable z.
To obtain the general solution in parametric form, let $z = a$ and solve by back-substitution: $x = -3 - a, y = 2 + 2a, z = a$; or $\mathbf{u} = (-3 - a, 2 + 2a, a)$.

(C) Reduce to echelon form. To eliminate x from the second and third equations, apply the operations $L_2 \rightarrow -3L_1 + L_2$ and $L_3 \rightarrow -5L_1 + L_3$ to obtain the equivalent system

$$\begin{array}{r} x + 2y - 3z = -1 \\ -7y + 11z = 10 \\ -7y + 11z = 7. \end{array}$$

The operation $L_3 \rightarrow -L_2 + L_3$ yields the degenerate equation $0 = -3$. Thus the system has no solution.

91. Reduce the system to echelon form. Apply the operations $L_2 \rightarrow -3L_1 + 2L_2$ and $L_3 \rightarrow -5L_1 + 2L_3$ and then $L_3 \rightarrow -5L_2 + L_3$ to obtain

$$\left.\begin{array}{r} 2x - 5y + 3z - 4s + 2t = 4 \\ y - 5z + 2s + 2t = 6 \\ 5y - 25z + 12s + 4t = 24 \end{array}\right\} \rightarrow \left\{\begin{array}{r} 2x - 5y + 3z - 4s + 2t = 4 \\ y - 5z + 2s + 2t = 6 \\ 2s - 6t = -6. \end{array}\right.$$

The system is now in echelon form. Solving for the leading unknowns, x, y, and s, in terms of the free variables, z and t, we obtain the free-variable form of the general solution:

$$x = 26 + 11z - 15t, \qquad y = 12 + 5z - 8t, \qquad s = -3 + 3t.$$

From this follows at once the parametric form

$$x = 26 + 11a - 15b, \qquad y = 12 + 5a - 8b, \qquad z = a, \qquad s = -3 + 3b, \qquad t = b.$$

92. Reduce the system to echelon form. Eliminate x from the second equation by the operation $\mathbf{L}_2 \rightarrow -\mathbf{L}_1 + \mathbf{L}_2$ and then eliminate y from the third equation by $\mathbf{L}_3 \rightarrow -k\mathbf{L}_2 + \mathbf{L}_3$; this yields

$$\left.\begin{aligned} x - 2y &= 1 \\ y + kz &= -3 \\ ky + 4z &= 6 \end{aligned}\right\} \rightarrow \left\{\begin{aligned} x - 2y &= 1 \\ y + kz &= -3 \\ (4 - k^2)z &= 6 + 3k. \end{aligned}\right.$$

The system has a unique solution if the coefficient of z in the third equation is not zero—that is, if $4 - k^2 \neq 0$. However, $4 - k^2 = 0$ if and only if $k = 2$ or $k = -2$. Hence the system has a unique solution if $k \neq 2$ and $k \neq -2$. If $k = 2$, then the third equation reduces to $0 = 12$, in which case the system has no solution. If $k = -2$, then the third equation becomes $0 = 0$, which can be deleted; the reduced system has a free variable z and hence an infinite number of solutions.

   (A) $k \neq 2$ and $k \neq -2$
   (B) $k = 2$
   (C) $k = -2$

93. $A\mathbf{x} = \mathbf{b}$, where $A = [a_{ij}]$ is the $m \times n$ coefficient matrix and $\mathbf{x}$ and $\mathbf{b}$ are the column vectors $\mathbf{x} = (x_1, x_2, \ldots, x_n)$ and $\mathbf{b} = (b_1, b_2, \ldots, b_n)$.
The augmented matrix of the system is the block matrix $[A \; \mathbf{b}]$.

94. $\begin{bmatrix} 2 & 3 & -4 \\ 1 & -2 & -5 \end{bmatrix} \begin{bmatrix} x \\ y \\ z \end{bmatrix} = \begin{bmatrix} 7 \\ 3 \end{bmatrix}$. Note that the size of the column of unknowns is not equal

to the size of the column of constants.

95. (A) Reduce the augmented matrix to echelon form and then to row canonical form:

$$\begin{bmatrix} 1 & -2 & -3 & 4 \\ 2 & -3 & 1 & 5 \end{bmatrix} \sim \begin{bmatrix} 1 & -2 & -3 & 4 \\ 0 & 1 & 7 & -3 \end{bmatrix} \sim \begin{bmatrix} 1 & 0 & 11 & -2 \\ 0 & 1 & 7 & -3 \end{bmatrix}.$$

Thus the free-variable form of the general solution is

$$\begin{aligned} x + 11z &= -2 \\ y + 7z &= -3 \end{aligned} \quad \text{or} \quad \begin{aligned} x &= -2 - 11z \\ y &= -3 - 7z. \end{aligned}$$

(Note that z is the free variable.)
   (B) Reduce the augmented matrix to echelon form and then to row canonical form:

$$\begin{bmatrix} 1 & 1 & -2 & 4 & 5 \\ 2 & 2 & -3 & 1 & 3 \\ 3 & 3 & -4 & -2 & 1 \end{bmatrix} \sim \begin{bmatrix} 1 & 1 & -2 & 4 & 5 \\ 0 & 0 & 1 & -7 & -7 \\ & & & & \end{bmatrix} \sim \begin{bmatrix} 1 & 1 & 0 & -10 & -9 \\ 0 & 0 & 1 & -7 & -7 \end{bmatrix}.$$

(The third row is deleted from the second matrix since it is a multiple of the second row and will result in a zero row.) Thus the free-variable form of the general solution of the system is as follows:

$$x + y - 10t = -9 \qquad \text{or} \qquad x = -9 - y + 10t$$
$$z - 7t = -7 \qquad \qquad z = -7 + 7t$$

Here the free variables are y and t.

(C) Reduce the augmented matrix to echelon form and then to row canonical form:

$$\begin{bmatrix} 1 & 2 & -3 & -2 & 4 & 1 \\ 2 & 5 & -8 & -1 & 6 & 4 \\ 1 & 4 & -7 & 5 & 2 & 8 \end{bmatrix} \sim \begin{bmatrix} 1 & 2 & -3 & -2 & 4 & 1 \\ 0 & 1 & -2 & 3 & -2 & 2 \\ 0 & 2 & -4 & 7 & -2 & 7 \end{bmatrix} \sim \begin{bmatrix} 1 & 2 & -3 & -2 & 4 & 1 \\ 0 & 1 & -2 & 3 & -2 & 2 \\ 0 & 0 & 0 & 1 & 2 & 3 \end{bmatrix}$$

$$\sim \begin{bmatrix} 1 & 2 & -3 & 0 & 8 & 7 \\ 0 & 1 & -2 & 0 & -8 & -7 \\ 0 & 0 & 0 & 1 & 2 & 3 \end{bmatrix} \sim \begin{bmatrix} 1 & 0 & 1 & 0 & 24 & 21 \\ 0 & 1 & -2 & 0 & -8 & -7 \\ 0 & 0 & 0 & 1 & 2 & 3 \end{bmatrix}.$$

Thus the free-variable form of the solution is

$$x + z + 24t = 21 \qquad \qquad x = 21 - z + 24t$$
$$y - 2z - 8t = -7 \qquad \text{or} \qquad y = -7 + 2z + 8t$$
$$s + 2t = 3 \qquad \qquad s = 3.$$

96. It can be shown that rank(A) equals the number of (nonzero) rows in an echelon form of A.

97. (A) Reduce the system to echelon form. Apply the operations $L_2 \to -2L_1 + L_2$ and $L_3 \to -3L_1 + L_3$ and then $L_3 \to -2L_2 + L_3$ to obtain

$$\left. \begin{array}{r} x + 3y - 2z + 5s - 3t = 0 \\ y + z - 3s + t = 0 \\ 2y + 2z - 5s = 0 \end{array} \right\} \to \left\{ \begin{array}{r} x + 3y - 2z + 5s - 3t = 0 \\ y + z - 3s + t = 0 \\ s - 2t = 0. \end{array} \right.$$

In echelon form, the system has two free variables, z and t; hence dim (W) = 2. A basis $\{u_1, u_2\}$ for W may be obtained as follows: (1) Set $z = 1$, $t = 0$. Back-substitution yields $s = 0$, then $y = -1$, and then $x = 5$. Thus, $u_1 = (5, -1, 1, 0, 0)$. (2) Set $z = 0$, $t = 1$. Back-substitution yields $s = 2$, then $y = 5$, and then $x = -2$. Thus, $u_2 = (-2, 5, 0, 2, 1)$.

(B) Reduce to echelon form. Apply the operations $L_2 \to -2L_1 + L_2$ and $L_3 \to -5L_2 + L_3$, and then $L_3 \to -2L_2 + L_3$ to obtain

$$\left. \begin{array}{r} x + 2y - 3z + 2s - 4t = 0 \\ z - 3s + 2t = 0 \\ 2z - 6s + 4t = 0 \end{array} \right\} \to \left\{ \begin{array}{r} x + 2y - 3z + 2s - 4t = 0 \\ z - 3s + 2t = 0. \end{array} \right.$$

In echelon form, the system has three free variables, y, s and t; hence dim $(W) = 3$. A basis $\{u_1, u_2, u_3\}$ for W is obtained as follows: (1) Set $y = 1$, $s = 0$, $t = 0$. Back-substitution yields the solution $u_1 = (-2, 2, 0, 0, 0)$. (2) Set $y = 0$, $s = 1$, $t = 0$. Back-substitution yields the solution $u_2 = (7, 0, 3, 1, 0)$. (3) Set $y = 0$, $s = 0$, $t = 1$. Back substitution yields the solution $u_3 = (-2, 0, -2, 0, 1)$.

98. $au_1 + bu_2 = a(5, -1, 1, 0, 0) + b(-2, 5, 0, 2, 1) = (5a - 2b, -a + 5b, a, 2b, b)$

99. Yes, since there are more unknowns than equations.

100. (A)
$$\begin{bmatrix} 1 \\ -6 \\ 5 \end{bmatrix} = \begin{bmatrix} x \\ 2x \\ 3x \end{bmatrix} + \begin{bmatrix} 2y \\ 5y \\ 8y \end{bmatrix} + \begin{bmatrix} 3z \\ 2z \\ 3z \end{bmatrix} = \begin{bmatrix} x + 2y + 3z \\ 2x + 5y + 2z \\ 3x + 8y + 3z \end{bmatrix}.$$

Set corresponding components of the vectors equal to each other, and reduce the system to echelon form:

$$\left. \begin{array}{r} x + 2y + 3z = 1 \\ 2x + 5y + 2z = -6 \\ 3x + 8y + 3z = 5 \end{array} \right\} \rightarrow \left\{ \begin{array}{r} x + 2y + 3z = 1 \\ y - 4z = -8 \\ 2y - 6z = 2 \end{array} \right. \rightarrow \left\{ \begin{array}{r} x + 2y + 3z = 2 \\ y - 4z = -8 \\ 2z = 18. \end{array} \right.$$

The system is triangular, and back-substitution yields the unique solution $x = -81$, $y = 28$, $z = 9$.

(B) Find the equivalent system of linear equations and solve. Writing

$$v = xu_1 + yu_2 + zu_3 = (x + y + 2z, x + 2y - z, x + 3y + z)$$

we obtain the system

$$\left. \begin{array}{r} x + y + 2z = 1 \\ x + 2y - z = -2 \\ x + 3y + z = 5 \end{array} \right\} \rightarrow \left\{ \begin{array}{r} x + y + 2z = 1 \\ y - 3z = -3 \\ 2y - z = 4 \end{array} \right. \rightarrow \left\{ \begin{array}{r} x + y + 2z = 1 \\ y - 3z = -3 \\ 5z = 10. \end{array} \right.$$

The unique solution of the triangular form is $x = -6$, $y = 3$, $z = 2$; thus $v = -6u_1 + 3u_2 + 2u_3$.

Chapter 4: Square Matrices

101. The diagonal consists of the elements from the upper left corner to the lower right corner of the matrix, and the trace is the sum of the diagonal element.
   (A) Diagonal: 1, 5, 9; trace: $1 + 5 + 9 = 15$.
   (B) Diagonal: $t - 2$, $t + 5$; trace: $2t + 3$.
   (C) The diagonal and trace are defined only for square matrices.

102. (A) $\text{tr}(I_2) = 1 + 1 = 2$ and $\text{tr}(I_3) = 1 + 1 + 1 = 3$.
   (B) I_n has n 1s on the diagonal; hence $\text{tr}(I_n) = n$.

103. (A) Note first that $I_m A$ is also an $m \times n$ matrix, say $I_m A = (f_{ij})$. But

$$f_{ij} = \sum_{k-1}^{m} \delta_{ik} a_{kj} = \delta_{ij} a_{ij} = a_{ij}.$$

Thus $I_m A = A$, since corresponding entries are equal.

(B) Note first that AI_n is also an $m \times n$ matrix, say $AI_n = (g_{ij})$. But

$$g_{ij} = \sum_{k-1}^{n} a_{ik} \delta_{kj} = a_{ij} \delta_{ij} = a_{ij}.$$

Thus $AI_n = A$, since corresponding entries are equal.

104. In each case, put 5s on the diagonal and 0s elsewhere:

$$\begin{bmatrix} 5 & 0 \\ 0 & 5 \end{bmatrix}, \begin{bmatrix} 5 & 0 & 0 \\ 0 & 5 & 0 \\ 0 & 0 & 5 \end{bmatrix}, \begin{bmatrix} 5 & & & \\ & 5 & & \\ & & 5 & \\ & & & 5 \end{bmatrix}.$$

(It is common practice to omit blocks or patterns of 0s as in the third matrix.)

105. Put the given scalars on the diagonal, with 0s elsewhere:

$$\begin{bmatrix} 3 & 0 & 0 \\ 0 & -7 & 0 \\ 0 & 0 & 2 \end{bmatrix}, \begin{bmatrix} 4 & 0 \\ 0 & -5 \end{bmatrix}, \begin{bmatrix} 6 & & & \\ & -3 & & \\ & & -9 & \\ & & & -1 \end{bmatrix}.$$

106. $AB = \begin{bmatrix} 5+12 & 4+22 \\ 15+24 & 12+44 \end{bmatrix} = \begin{bmatrix} 17 & 26 \\ 39 & 56 \end{bmatrix}$ and $BA = \begin{bmatrix} 5+12 & 10+16 \\ 6+33 & 12+44 \end{bmatrix} = \begin{bmatrix} 17 & 26 \\ 39 & 56 \end{bmatrix}.$

Since $AB = BA$, the matrices commute.

107. (A) $A^2 = AA = \begin{bmatrix} 1 & 2 \\ 4 & -3 \end{bmatrix} \begin{bmatrix} 1 & 2 \\ 4 & -3 \end{bmatrix} = \begin{bmatrix} 1+8 & 2-6 \\ 4-12 & 8+9 \end{bmatrix} = \begin{bmatrix} 9 & -4 \\ -8 & 17 \end{bmatrix}$

(B) $A^3 = AA^2 = \begin{bmatrix} 1 & 2 \\ 4 & -3 \end{bmatrix} \begin{bmatrix} 9 & -4 \\ -8 & 17 \end{bmatrix} = \begin{bmatrix} 9-16 & -4+34 \\ 36+24 & -16-51 \end{bmatrix} = \begin{bmatrix} -7 & 30 \\ 60 & -67 \end{bmatrix}$

108. $f(A) = 2A^3 - 4A + 5I = 2 \begin{bmatrix} -7 & 30 \\ 60 & -67 \end{bmatrix} - 4 \begin{bmatrix} 1 & 2 \\ 4 & -3 \end{bmatrix} + 5 \begin{bmatrix} 1 & 0 \\ 0 & 1 \end{bmatrix}$

$$= \begin{bmatrix} -14 & 60 \\ 120 & -134 \end{bmatrix} + \begin{bmatrix} -4 & -8 \\ -16 & 12 \end{bmatrix} + \begin{bmatrix} 5 & 0 \\ 0 & 5 \end{bmatrix}$$

$$= \begin{bmatrix} -14-4+5 & 60-8+0 \\ 120-16+0 & -134+12+5 \end{bmatrix} = \begin{bmatrix} -13 & 52 \\ 104 & -117 \end{bmatrix}$$

109. $g(A) = A^2 + 2A - 11I = \begin{bmatrix} 9 & -4 \\ -8 & 17 \end{bmatrix} + 2\begin{bmatrix} 1 & 2 \\ 4 & -3 \end{bmatrix} - 11\begin{bmatrix} 1 & 0 \\ 0 & 1 \end{bmatrix}$

$$= \begin{bmatrix} 9 & -4 \\ -8 & 17 \end{bmatrix} + \begin{bmatrix} 2 & 4 \\ 8 & -6 \end{bmatrix} + \begin{bmatrix} -11 & 0 \\ 0 & -11 \end{bmatrix}$$

$$= \begin{bmatrix} 9+2-11 & -4+4+0 \\ -8+8+0 & 17-6-11 \end{bmatrix} = \begin{bmatrix} 0 & 0 \\ 0 & 0 \end{bmatrix}$$

(To the interested reader: Look up the *Cayley–Hamilton theorem*.)

110. $AB = \begin{bmatrix} -11+0+12 & 2+0-2 & 2+0-2 \\ -22+4+18 & 4+0-3 & 4-1-3 \\ -44-4+48 & 8+0-8 & 8+1-8 \end{bmatrix} = \begin{bmatrix} 1 & 0 & 0 \\ 0 & 1 & 0 \\ 0 & 0 & 1 \end{bmatrix} = I.$

By Theorem 4.1, $AB = I$ if and only if $BA = I$; hence we do not need to test $BA = I$.

111. Use the formula in Remark 4.1. That is, first find $|A|$. If $|A| \neq 0$ then (i) interchange the diagonal elements, (ii) negate the other elements, and (iii) multiply by $1/|A|$.

    (A) $|A| = (3)(3) - (5)(2) = -1 \neq 0$. Hence

$$A^{-1} = -1\begin{bmatrix} 3 & -5 \\ -2 & 3 \end{bmatrix} = \begin{bmatrix} -3 & 5 \\ 2 & -3 \end{bmatrix}.$$

    (B) $|A| = (5)(2) - (3)(4) = -2 \neq 0$. Hence

$$A^{-1} = -\frac{1}{2}\begin{bmatrix} 2 & -3 \\ -4 & 5 \end{bmatrix} = \begin{bmatrix} -1 & \frac{3}{2} \\ 2 & \frac{-5}{2} \end{bmatrix}.$$

    (C) $|A| = (-2)(-9) - (6)(3) = 0$; hence A has no inverse.

112. (A) Form the block matrix $M = (A : I)$ and reduce M to echelon form:

$$M = \begin{bmatrix} 1 & 0 & 2 & \vdots & 1 & 0 & 0 \\ 2 & -1 & 3 & \vdots & 0 & 1 & 0 \\ 4 & 1 & 8 & \vdots & 0 & 0 & 1 \end{bmatrix} \sim \begin{bmatrix} 1 & 0 & 2 & \vdots & 1 & 0 & 0 \\ 0 & -1 & -1 & \vdots & -2 & 1 & 0 \\ 0 & 1 & 0 & \vdots & -4 & 0 & 1 \end{bmatrix}$$

$$\sim \begin{bmatrix} 1 & 0 & 2 & \vdots & 1 & 0 & 0 \\ 0 & -1 & -1 & \vdots & -2 & 1 & 0 \\ 0 & 0 & -1 & \vdots & -6 & 1 & 1 \end{bmatrix}.$$

In echelon form, the left half of M is in triangular form; hence A is invertible. Further row reduce M to row canonical form:

$$M \sim \begin{bmatrix} 1 & 0 & 0 & \vdots & -11 & 2 & 2 \\ 0 & -1 & 0 & \vdots & 4 & 0 & -1 \\ 0 & 0 & 1 & \vdots & 6 & -1 & -1 \end{bmatrix} \sim \begin{bmatrix} 1 & 0 & 0 & \vdots & -11 & 2 & 2 \\ 0 & 1 & 0 & \vdots & -4 & 0 & 1 \\ 0 & 0 & 1 & \vdots & 6 & -1 & -1 \end{bmatrix}.$$

The final block matrix is in the form $(I : A^{-1})$.

(B) Form the block matrix $M = (B : I)$ and reduce M to echelon form:

$$M = \begin{bmatrix} 1 & -2 & 2 & | & 1 & 0 & 0 \\ 2 & -3 & 6 & | & 0 & 1 & 0 \\ 1 & 1 & 7 & | & 0 & 0 & 1 \end{bmatrix} \sim \begin{bmatrix} 1 & -2 & 2 & | & 1 & 0 & 0 \\ 0 & 1 & 2 & | & -2 & 1 & 0 \\ 0 & 3 & 5 & | & -1 & 0 & 1 \end{bmatrix}$$

$$\sim \begin{bmatrix} 1 & -2 & 2 & | & 1 & 0 & 0 \\ 0 & 1 & 2 & | & -2 & 1 & 0 \\ 0 & 0 & -1 & | & 5 & -3 & 1 \end{bmatrix}.$$

In echelon form, the left half of M is in triangular form; hence B has an inverse. Further row reduce M to row canonical form:

$$M \sim \begin{bmatrix} 1 & -2 & 0 & | & 11 & -6 & 2 \\ 0 & 1 & 0 & | & 8 & -5 & 2 \\ 0 & 0 & 1 & | & -5 & 3 & -1 \end{bmatrix} \sim \begin{bmatrix} 1 & 0 & 0 & | & 27 & -16 & 6 \\ 0 & 1 & 0 & | & 8 & -5 & 2 \\ 0 & 0 & 1 & | & -5 & 3 & -1 \end{bmatrix}.$$

The final matrix has the form $(I : B^{-1})$.

113. Form the block matrix $M = (B : I)$ and row reduce to echelon form:

$$\begin{bmatrix} 1 & 3 & -4 & | & 1 & 0 & 0 \\ 1 & 5 & -1 & | & 0 & 1 & 0 \\ 3 & 13 & -6 & | & 0 & 0 & 1 \end{bmatrix} \sim \begin{bmatrix} 1 & 3 & -4 & | & 1 & 0 & 0 \\ 0 & 2 & 3 & | & -1 & 1 & 0 \\ 0 & 4 & 6 & | & -3 & 0 & 1 \end{bmatrix} \sim \begin{bmatrix} 1 & 3 & -4 & | & 1 & 0 & 0 \\ 0 & 2 & 3 & | & -1 & 1 & 0 \\ 0 & 0 & 0 & | & -1 & -2 & 1 \end{bmatrix}.$$

In echelon form, M has a zero row in its left half; that is, B is not row reducible to triangular form. Accordingly, B is not invertible.

114. Choose $B = (-1)A$. Then $A + B = 0$, which is not invertible.

115. (A) Apply the operation $R_1 \leftrightarrow R_2$ to I_3; that is, interchange the first and second rows to obtain

$$E_1 = \begin{bmatrix} 0 & 1 & 0 \\ 1 & 0 & 0 \\ 0 & 0 & 1 \end{bmatrix}.$$

(B) Apply the operation $R_3 \rightarrow -7R_3$ to I_3; that is, multiply the third row of I_3 by -7 to obtain

$$E_2 = \begin{bmatrix} 1 & 0 & 0 \\ 0 & 1 & 0 \\ 0 & 0 & -7 \end{bmatrix}.$$

(C) Apply the operation $R_2 \rightarrow -3R_1 + R_2$ to I_3; that is, replace the second row of I_3 by $-3R_1 + R_2$ to obtain

$$E_3 = \begin{bmatrix} 1 & 0 & 0 \\ -3 & 1 & 0 \\ 0 & 0 & 1 \end{bmatrix}.$$

116. Let E be the elementary matrix corresponding to the elementary row operation e: $e(I) = E$. Let e' be the inverse operation of e and let E' be its corresponding elementary matrix. Then $I = e'$ $(e(I)) = e'(E) = E'E$ and $I = e(e'(I)) = e(E') = EE$. Therefore E' is the inverse of E.

117. Let E_i be the elementary matrix corresponding to the operation e_i. Then, by hypothesis, $E_n \ldots E_2E_1A = I$. Thus $(E_n \ldots E_2E_1I)A = I$ and hence $A^{-1} = E_n \ldots E_2E_1I$. In other words, A^{-1} can be obtained from I by applying the elementary row operations $e_1, \ldots, e_n$.

118. If AB is invertible, then there exists a matrix C such that $(AB)C = I$. Hence $A(BC) = I$ and BC is the inverse of A.

119. (A) Let $A + B = [c_{ij}]$. If $i > j$, then $c_{ij} = a_{ij} + b_{ij} = 0 + 0 = 0$. Hence $A + B$ is upper triangular. Also, the $c_{ij} = a_{ij} + b_{ij}$ are the diagonal elements.
(B) Let $kA = [c_{ij}]$. If $i > j$, then $c_i = ka_{ij} = k \cdot 0 = 0$. Hence kA is upper triangular. Also, the $c_{ij} = ka_{ij}$ are the diagonal elements.

120. Let $AB = [c_{ij}]$. Then $c_{ij} = \sum_{k=1}^{n} a_{ik}a_{kj}$.
(A) If $i > j$, then, for any k, either $i > k$ or $k > j$, so either $a_{ik} = 0$ or $b_{kj} = 0$. Thus $c_{ik} = 0$, and AB is upper triangular.
(B) Here $c_{ii} = \sum_{k=1}^{n} a_{ik}a_{ki}$.

But for $k < i$, $a_{ik} = 0$; and for $k > i$, $b_{ki} = 0$. Hence $c_{ii} = a_{ii}b_{ii}$, as claimed.

121. (A) Diagonal matrices must have 0s off the diagonal:

$$\begin{bmatrix} 1 & 0 \\ 0 & 1 \end{bmatrix}, \begin{bmatrix} 1 & 0 \\ 0 & 0 \end{bmatrix}, \begin{bmatrix} 0 & 0 \\ 0 & 1 \end{bmatrix}, \begin{bmatrix} 0 & 0 \\ 0 & 0 \end{bmatrix}.$$

(B) Upper triangular matrices must have 0s below the diagonal. This gives the four matrices of (a), plus the four matrices obtained from them by changing the (1, 2)-element to 1.

122. If $A = [a_{ij}]$ is skew-symmetric, then $a_{ii} = -a_{ii}$. Hence each $a_{ii} = 0$.

123. (A) By inspection, $A^T = -A$; thus A is skew-symmetric.
(B) By inspection, $B^T = B$; thus B is symmetric.
(C) Since C is not square, C is neither symmetric nor skew-symmetric.

124. Set the symmetric elements (mirror images in the diagonal) $x + 2$ and $2x - 3$ equal to each other, to obtain $x = 5$; hence $A = \begin{bmatrix} 4 & 7 \\ 7 & 6 \end{bmatrix}$.

125. Calculate

$$A^{\mathrm{T}} = \begin{bmatrix} 2 & 7 \\ 3 & 8 \end{bmatrix}, \quad A + A^{\mathrm{T}} = \begin{bmatrix} 4 & 10 \\ 10 & 16 \end{bmatrix}, \quad A - A^{\mathrm{T}} = \begin{bmatrix} 0 & -4 \\ 4 & 0 \end{bmatrix}.$$

Then

$$B = \tfrac{1}{2}(A + A^{\mathrm{T}}) = \begin{bmatrix} 2 & 5 \\ 5 & 8 \end{bmatrix}, \quad C = \tfrac{1}{2}(A - A^{\mathrm{T}}) = \begin{bmatrix} 0 & -2 \\ 2 & 0 \end{bmatrix}.$$

126. The vectors $\mathbf{u}_1, \mathbf{u}_2, \ldots, \mathbf{u}_r$ form an orthonormal set if the vectors are pairwise orthogonal ($\mathbf{u}_i \cdot \mathbf{u}_j = 0$ for $i \neq j$) and if the vectors have unit lengths ($\mathbf{u}_i \cdot \mathbf{u}_j = 1$ for $i = 1, 2, \ldots, r$). In terms of the Kronecker delta, the condition for orthogonality is $\mathbf{u}_i \cdot \mathbf{u}_j = \delta_{ij}$.

127. If A is orthogonal, then

$$AA^{\mathrm{T}} = \begin{bmatrix} a_1 & a_2 & a_3 \\ b_1 & b_2 & b_3 \\ c_1 & c_2 & c_3 \end{bmatrix} \begin{bmatrix} a_1 & b_1 & c_1 \\ a_2 & b_2 & c_2 \\ a_3 & b_3 & c_3 \end{bmatrix} = \begin{bmatrix} 1 & 0 & 0 \\ 0 & 1 & 0 \\ 0 & 0 & 1 \end{bmatrix} = I.$$

This yields

$$a_1^2 + a_2^2 + a_3^2 = \mathbf{u}_1 \cdot \mathbf{u}_1 = 1 \qquad a_1 b_1 + a_2 b_2 + a_3 b_3 = \mathbf{u}_1 \cdot \mathbf{u}_2 = 0 \qquad a_1 c_1 + a_2 c_2 + a_3 c_3 = \mathbf{u}_1 \cdot \mathbf{u}_3 = 0$$

$$b_1 a_1 + b_2 a_2 + b_3 a_3 = \mathbf{u}_2 \cdot \mathbf{u}_1 = 0 \qquad b_1^2 + b_2^2 + b_3^2 = \mathbf{u}_2 \cdot \mathbf{u}_2 = 1 \qquad b_1 c_1 + b_2 c_2 + b_3 c_3 = \mathbf{u}_2 \cdot \mathbf{u}_3 = 0$$

$$c_1 a_1 + c_2 a_2 + c_3 a_3 = \mathbf{u}_3 \cdot \mathbf{u}_1 = 0 \qquad c_1 b_1 + c_2 b_2 + c_3 b_3 = \mathbf{u}_3 \cdot \mathbf{u}_2 = 0 \qquad c_1^2 + c_2^2 + c_3^2 = \mathbf{u}_3 \cdot \mathbf{u}_3 = 1$$

that is, $\mathbf{u}_i \cdot \mathbf{u}_j = \delta_{ij}$. Accordingly, $\mathbf{u}_1, \mathbf{u}_2, \mathbf{u}_3$ form an orthonormal set. The converse follows from the fact that each step is reversible.

128. Let R_1, R_2 and C_1, C_2 denote, respectively, the rows and columns of A. Since $R_1 \cdot R_2 = 0$, we get $x/\sqrt{5} + 2y/\sqrt{5} = 0$, or $x + 2y = 0$. Since C_1 is a unit vector, we get $x^2 + 1/5 = 1$, or $x = \pm 2/\sqrt{5}$.

Case (i): $x = 2/\sqrt{5}$. Then $x + 2y = 0$ yields $y = -1/\sqrt{5}$.
Case (ii): $x = -2/\sqrt{5}$. Then $x + 2y = 0$ yields $y = 1/\sqrt{5}$.

In other words, there are exactly two possibilities:

$$A = \begin{bmatrix} 1/\sqrt{5} & 2/\sqrt{5} \\ -2/\sqrt{5} & 1/\sqrt{5} \end{bmatrix} \quad \text{and} \quad A = \begin{bmatrix} 1/\sqrt{5} & 2/\sqrt{5} \\ 2/\sqrt{5} & -1/\sqrt{5} \end{bmatrix}.$$

129. Let $A = \begin{bmatrix} a & b \\ c & d \end{bmatrix}$. Then a, b, c, d are real numbers, and the rows of A form an orthonormal set. Hence

$$a^2 + b^2 = 1, \; c^2 + d^2 = 1, \; ac + bd = 0.$$

Similarly, the columns form an orthonormal set, so

$$a^2 + c^2 = 1, \; b^2 + d^2 = 1, \; ab + cd = 0.$$

Therefore, $c^2 = 1 - a^2 = b^2$, whence $c = \pm b$.
Case (i): $c = +b$. Then $b(a + d) = 0$, or $d = -a$; the corresponding matrix is $\begin{bmatrix} a & b \\ b & -a \end{bmatrix}$.

Case (ii): $c = -b$. Then $b(d - a) = 0$, or $d = a$; the corresponding matrix is $\begin{bmatrix} a & b \\ -b & a \end{bmatrix}$.

130. Let a and b be any real numbers such that $a^2 + b^2 = 1$. Then there exists a real number θ such that $a = \cos \theta$ and $b = \sin \theta$.

131. (A) No. Although A is a 5×5 square matrix and is a 3×3 block matrix, the second and third diagonal blocks are not square matrices.
  (B) Yes.
  (C) One horizontal line is between the second and third rows; hence add a vertical line between the second and third columns. The other horizontal line is between the fourth and fifth rows; hence add a vertical line between the fourth and fifth columns. (The horizontal lines and the vertical lines must be symmetrically placed to obtain a square block matrix.) This yields the square block matrix

$$C = \left[\begin{array}{cc:cc:c} 1 & 2 & 3 & 4 & 5 \\ 1 & 1 & 1 & 1 & 1 \\ \hdashline 9 & 8 & 7 & 6 & 5 \\ 3 & 3 & 3 & 3 & 3 \\ \hdashline 1 & 3 & 5 & 7 & 9 \end{array} \right].$$

132. $A = \left[\begin{array}{c:cc} 1 & 0 & 0 \\ \hdashline 0 & 0 & 2 \\ 0 & 0 & 3 \end{array} \right]$

133. $B = \left[\begin{array}{cc:cc:c} 1 & 2 & 0 & 0 & 0 \\ 3 & 0 & 0 & 0 & 0 \\ \hdashline 0 & 0 & 4 & 0 & 0 \\ 0 & 0 & 5 & 0 & 0 \\ \hdashline 0 & 0 & 0 & 0 & 6 \end{array} \right]$

134. Considered as a single 3×3 block, C (or any other 3×3 matrix) is a block diagonal matrix; no further partitioning of C is possible.

135. (A) Simply add the diagonal blocks: $M + N = \text{diag}(A_1 + B_1, A_2 + B_2, \ldots, A_r + B_r)$.
   (B) Simply multiply the diagonal blocks by k: $kM = \text{diag}(kA_1, kA_2, \ldots, kA_r)$.
   (C) Simply multiply corresponding diagonal blocks: $MN = \text{diag}(A_1B_1, A_2B_2, \ldots, A_rB_r)$.
   (D) Find $f(A_i)$ for each diagonal block. Then $f(M) = \text{diag}[\,f(A_1), f(A_2), \ldots, f(a_r)]$.

Chapter 5: Determinants

136. For any n-square matrices A and B, $\det(AB) = \det(A)\det(B)$.

137. $24, -6, t + 2$

138. (A) $\begin{vmatrix} 5 & 4 \\ 2 & 3 \end{vmatrix} = (5)(3) - (4)(2) = 15 - 8 = 7$

   (B) $\begin{vmatrix} 2 & 1 \\ -4 & 6 \end{vmatrix} = (2)(6) - (1)(-4) = 12 + 4 = 16$

   (C) $\begin{vmatrix} 3 & -2 \\ 4 & 5 \end{vmatrix} = 15 + 8 = 23$

   (D) $\begin{vmatrix} 4 & -5 \\ -1 & -2 \end{vmatrix} = -8 - 5 = -13$

   (E) $\begin{vmatrix} a & b \\ c & d \end{vmatrix} = ad - bc$

139. $\begin{vmatrix} t-2 & 3 \\ 4 & t-1 \end{vmatrix} = t^2 - 3t + 2 - 12 = t^2 - 3t - 10 = 0$, or $(t-5)(t+2) = 0$.

Hence, $t = 5$ or $t = -2$.

140. (A) Use Figure 5.1,

$\begin{vmatrix} 2 & 1 & 1 \\ 0 & 5 & -2 \\ 1 & -3 & 4 \end{vmatrix} = (2)(5)(4) + (1)(-2)(1) + (1)(-3)(0) - (1)(5)(1) - (-3)(-2)(2) - (4)(1)(0)$

$= 40 - 2 + 0 - 5 - 12 - 0 = 21$

   (B) $\begin{vmatrix} 3 & -2 & -4 \\ 2 & 5 & -1 \\ 0 & 6 & 1 \end{vmatrix} = (3)(5)(1) + (-2)(-1)(0) + (-4)(6)(2) - (0)(5)(-4) - (6)(-1)(3) - (1)(-2)(2)$

$= 15 + 0 - 48 - 0 + 18 + 4 = -11$

   (C) $\begin{vmatrix} -2 & -1 & 4 \\ 6 & -3 & -2 \\ 4 & 1 & 2 \end{vmatrix} = 12 + 8 + 24 + 48 - 4 + 12 = 100$

141. (A) Expand by the first row:

$$\det(A) = \begin{vmatrix} 1 & 2 & 3 \\ 4 & -2 & 3 \\ 0 & 5 & -1 \end{vmatrix} = \begin{vmatrix} \boxed{1} & 2 & 3 \\ 4 & -2 & 3 \\ 0 & 5 & -1 \end{vmatrix} -2 \begin{vmatrix} 1 & \boxed{2} & 3 \\ 4 & -2 & 3 \\ 0 & 5 & -1 \end{vmatrix} +3 \begin{vmatrix} 1 & 2 & \boxed{3} \\ 4 & -2 & 3 \\ 0 & 5 & -1 \end{vmatrix}$$

$$= 1 \begin{vmatrix} -2 & 3 \\ 5 & -1 \end{vmatrix} -2 \begin{vmatrix} 4 & 3 \\ 0 & -1 \end{vmatrix} +3 \begin{vmatrix} 4 & -2 \\ 0 & 5 \end{vmatrix}$$

$$= 1(2-15)-2(-4+0)+3(20+0) = -13+8+60 = 55$$

(B) $\det(B) = \begin{vmatrix} 2 & 0 & 1 \\ 3 & 2 & -3 \\ -1 & -3 & 5 \end{vmatrix} = 2(10-9)+1(-9+2) = -5$

(C) $\det(C) = \begin{vmatrix} 1 & 0 & 0 \\ 3 & 2 & -4 \\ 4 & 1 & 3 \end{vmatrix} = 1(6+4) = 10$

(The calculations become more and more simple as there are more and more 0s in the expanding row.)

142. First put the system in standard form with the unknowns appearing in columns:

$$2x + 3y - z = 1$$
$$3x + 5y + 2z = 8$$
$$x - 2y - 3z = -1.$$

Compute the determinant D of the matrix of coefficients:

$$D = \begin{vmatrix} 2 & 3 & -1 \\ 3 & 5 & 2 \\ 1 & -2 & -3 \end{vmatrix} = -30+6+6+5+8+27 = 22.$$

Since $D \neq 0$, the system has a unique solution. To compute N_x, N_y, and N_z, replace the coefficients of x, y, and z in the matrix of coefficients by the constant terms:

$$N_x = \begin{vmatrix} 1 & 3 & -1 \\ 8 & 5 & 2 \\ -1 & -2 & -3 \end{vmatrix} = -15-6+16-5+4+72 = 66$$

$$N_y = \begin{vmatrix} 2 & 1 & -1 \\ 3 & 8 & 2 \\ 1 & -1 & -3 \end{vmatrix} = -48+2+3+8+4+9 = -22$$

$$N_z = \begin{vmatrix} 2 & 3 & 1 \\ 3 & 5 & 8 \\ 1 & -2 & -1 \end{vmatrix} = -10+24-6-5+32+9 = 44.$$

Hence $x = N_x/D = 3$, $\quad y = N_y/D = -1$, $\quad z = N_z/D = 2$.

143. There are $2! = 2 \cdot 1 = 2$ permutations in S_2: 12 and 21.

144. There are $3! = 3 \cdot 2 \cdot 1 = 6$ permutations in S_3: 123, 132, 213, 231, 312, 321.

145. For each element, count the number of elements smaller than it and to the right of it. Thus, 3 produces the inversions (3, 1) and (3, 2); 5 produces the inversions (5, 1), (5, 4), and (5, 2); 1 produces no inversion; 4 produces the inversion (4, 2); 2 produces no inversion. Since there are, in all, six inversions, σ is even and sign $\sigma = 1$.

146. The identity permutation $\varepsilon = 123...n$ is even because there are no inversions in ε.

147. (A) The permutation 12 is even, and the permutations 21 is odd.
   (B) The permutations 123, 231, and 312 are even, and the permutations 132, 213, and 321 are odd.

148. $P^{-1}P = I$. Hence $1 = |I| = |P^{-1}P| = |P^{-1}||P|$, and so $|P^{-1}| = |P|^{-1}$.

149. $|B| = |P^{-1}AP| = |P^{-1}||A||P| = |A||P^{-1}||P| = |A|$. (Although the matrices P^{-1} and A need not commute, their determinants, as scalars, do commute.)

150. (A) Delete the second row and the first column of A:

$$M_{21} = \begin{bmatrix} \boxed{2 & 3 & 4} \\ 5 & 6 & 7 \\ \boxed{8 & 9 & 1} \end{bmatrix} = \begin{bmatrix} 3 & 4 \\ 9 & 1 \end{bmatrix} \qquad \text{and so} \qquad |M_{21}| = 3 - 36 = -33.$$

   (B) Multiply the minor $|M_{21}|$ by the sign $(-1)^{2+1} = -1$; that is, $A_{21} = (-1)(-33) = 33$.
   (C) Delete the second row and second column from A and then find the determinant:

$$|M_{21}| = \begin{vmatrix} 2 & 3 & 4 \\ 5 & 6 & 7 \\ 8 & 9 & 1 \end{vmatrix} = \begin{vmatrix} 2 & 4 \\ 8 & 1 \end{vmatrix} = 2 - 32 = -30.$$

   (D) Multiply the minor $|M_{22}|$ by the appropriate sign:
$A_{22} = (-1)^{2+2}|M_{22}| = (+1)(-30) = -30$.
   (E) Delete second row and third column from A and find its determinant:

$$|M_{23}| = \begin{vmatrix} 2 & 3 \\ 8 & 9 \end{vmatrix} = 18 - 24 = -6.$$

   (F) $A_{23} = (-1)^{2+3}|M_{23}| = (-1)(-6) = 6$

151. (A) Use $a_{31} = 1$ as a pivot and apply the row operations $R_1 \rightarrow -2R_3 + R_1$, $R_2 \rightarrow 2R_3 + R_2$, and $R_4 \rightarrow R_3 + R_4$:

$$|A| = \begin{vmatrix} 2 & 5 & -3 & -2 \\ -2 & -3 & 2 & -5 \\ 1 & 3 & -2 & 2 \\ -1 & -6 & 4 & 3 \end{vmatrix} = \begin{vmatrix} 0 & -1 & 1 & -6 \\ 0 & 3 & -2 & -1 \\ 1 & 3 & -2 & 2 \\ 0 & -3 & 2 & 5 \end{vmatrix} = + \begin{vmatrix} -1 & 1 & -6 \\ 3 & -2 & -1 \\ -3 & 2 & 5 \end{vmatrix}$$

$$= 10 + 3 - 36 + 36 - 2 - 15 = -4.$$

(B) Use $b_{21} = 1$ as a pivot and put 0s in the other entries in the second row by the column operations $C_2 \rightarrow 2C_1 + C_2$, $C_3 \rightarrow 2C_1 + C_3$, and $C_4 \rightarrow -3C_1 + C_4$. Then

$$|B| = \begin{vmatrix} 3 & -2 & -5 & 4 \\ 1 & -2 & -2 & 3 \\ -2 & 4 & 7 & -3 \\ 2 & -3 & -5 & 8 \end{vmatrix} = \begin{vmatrix} 3 & -2+2(3) & -5+2(3) & 4-3(3) \\ 1 & -2+2(1) & -2+2(1) & 3-3(1) \\ -2 & 4+2(-2) & 7+2(-2) & -3-3(-2) \\ 2 & -3+2(2) & -5+2(2) & 8-3(2) \end{vmatrix}$$

$$= \begin{vmatrix} 3 & 4 & 1 & -5 \\ 1 & 0 & 0 & 0 \\ -2 & 0 & 3 & 3 \\ 2 & 1 & -1 & 2 \end{vmatrix} = - \begin{vmatrix} 4 & 1 & -5 \\ 0 & 3 & 3 \\ 1 & -1 & 2 \end{vmatrix} = -(24 + 3 + 0 + 15 + 12 - 0) = -54.$$

152. First find the nine cofactors A_{ij} of A:

$$A_{11} = + \begin{vmatrix} 6 & 7 \\ 9 & 1 \end{vmatrix} = -57, \quad A_{12} = - \begin{vmatrix} 5 & 7 \\ 8 & 1 \end{vmatrix} = 51, \quad A_{13} = + \begin{vmatrix} 5 & 6 \\ 8 & 9 \end{vmatrix} = -3,$$

$$A_{21} = - \begin{vmatrix} 3 & 4 \\ 9 & 1 \end{vmatrix} = 33, \quad A_{22} = + \begin{vmatrix} 2 & 4 \\ 8 & 1 \end{vmatrix} = -30, \quad A_{23} = - \begin{vmatrix} 2 & 3 \\ 8 & 9 \end{vmatrix} = 6,$$

$$A_{31} = + \begin{vmatrix} 3 & 4 \\ 6 & 7 \end{vmatrix} = -3, \quad A_{32} = - \begin{vmatrix} 2 & 4 \\ 5 & 7 \end{vmatrix} = 6, \quad A_{33} = + \begin{vmatrix} 2 & 3 \\ 5 & 6 \end{vmatrix} = -3.$$

Take the transpose of the above matrix of cofactors:

$$\text{adj}(A) = \begin{bmatrix} -57 & 33 & -3 \\ 51 & -30 & 6 \\ -3 & 6 & -3 \end{bmatrix}.$$

153. (A) $|A| = -40 + 6 + 0 - 16 + 4 + 0 = -46.$

(B) First find the nine cofactors A_{ij} of A:

$$A_{11} = + \begin{vmatrix} -4 & 2 \\ -1 & 5 \end{vmatrix} = -18, \quad A_{12} = - \begin{vmatrix} 0 & 2 \\ 1 & 5 \end{vmatrix} = 2, \quad A_{13} = + \begin{vmatrix} 0 & -4 \\ 1 & -1 \end{vmatrix} = 4,$$

$$A_{21} = - \begin{vmatrix} 3 & -4 \\ -1 & 5 \end{vmatrix} = -11, \quad A_{22} = + \begin{vmatrix} 2 & -4 \\ 1 & 5 \end{vmatrix} = 14, \quad A_{23} = - \begin{vmatrix} 2 & 3 \\ 1 & -1 \end{vmatrix} = 5,$$

$$A_{31} = + \begin{vmatrix} 3 & -4 \\ -4 & 2 \end{vmatrix} = -10, \quad A_{32} = - \begin{vmatrix} 2 & -4 \\ 0 & 2 \end{vmatrix} = -4, \quad A_{33} = + \begin{vmatrix} 2 & 3 \\ 0 & -4 \end{vmatrix} = -8.$$

Then adj(A) is the transpose of the matrix of cofactors:

$$\text{adj}(A) = \begin{bmatrix} -18 & -11 & -10 \\ 2 & 14 & -4 \\ 4 & 5 & -8 \end{bmatrix}.$$

(C) $A \cdot [\text{adj}(A)] = \begin{bmatrix} 2 & 3 & -4 \\ 0 & -4 & 2 \\ 1 & -1 & 5 \end{bmatrix} \begin{bmatrix} -18 & -11 & -10 \\ 2 & 14 & -4 \\ 4 & 5 & -8 \end{bmatrix}$

$$= \begin{bmatrix} -46 & 0 & 0 \\ 0 & -46 & 0 \\ 0 & 0 & -46 \end{bmatrix} = -46 \begin{bmatrix} 1 & 0 & 0 \\ 0 & 1 & 0 \\ 0 & 0 & 1 \end{bmatrix} = -46I = |A|I.$$

(D) Since $|A| \neq 0$,

$$A^{-1} = \frac{1}{|A|} \text{adj}(A) \begin{bmatrix} -18/-46 & -111/-46 & -10/-46 \\ 2/-46 & 14/-46 & -4/-46 \\ 4/-46 & 5/-46 & -8/-46 \end{bmatrix} = \begin{bmatrix} 9/23 & 11/46 & 5/23 \\ -1/23 & -7/23 & 2/23 \\ -2/23 & -5/46 & 4/23 \end{bmatrix}.$$

154. Compute

$$D = |A| = \begin{vmatrix} 1 & 1 & 1 & 1 \\ 1 & 2 & 3 & 4 \\ 2 & 3 & 5 & 9 \\ 1 & 1 & 2 & 7 \end{vmatrix} = 2, \quad N_1 = |A_1| = \begin{vmatrix} 2 & 1 & 1 & 1 \\ 2 & 2 & 3 & 4 \\ 2 & 3 & 5 & 9 \\ 2 & 1 & 2 & 7 \end{vmatrix} = -4,$$

$$N_2 = \begin{vmatrix} 1 & 2 & 1 & 1 \\ 1 & 2 & 3 & 4 \\ 2 & 2 & 5 & 9 \\ 1 & 2 & 2 & 7 \end{vmatrix} = 18, \quad N_3 = \begin{vmatrix} 1 & 1 & 2 & 1 \\ 1 & 2 & 2 & 4 \\ 2 & 3 & 2 & 9 \\ 1 & 1 & 2 & 7 \end{vmatrix} = -12, \quad N_4 = \begin{vmatrix} 1 & 1 & 1 & 2 \\ 1 & 2 & 3 & 2 \\ 2 & 3 & 5 & 2 \\ 1 & 1 & 2 & 2 \end{vmatrix} = 2.$$

Then $x_1 = N_1/D = -2$, $x_2 = N_2/D = 9$, $x_3 = N_3/D = -6$, $x_4 = N_4/D = 1$.

155. Note that M is an upper triangular block matrix. Evaluate the determinant of each diagonal block:

$$\begin{vmatrix} 2 & 3 \\ -1 & 5 \end{vmatrix} = 10 + 3 = 13, \quad \begin{vmatrix} 2 & 1 & 5 \\ 3 & -1 & 4 \\ 5 & 2 & 6 \end{vmatrix} = -12 + 20 + 30 + 25 - 16 - 18 = 29.$$

Then $|M| = (13)(29) = 377$.

156. The subscripts 3 and 5 refer to the rows of A, and the superscripts 1 and 4 refer to the columns of A. Hence

$$\left| A_{3,5}^{1,4} \right| = \begin{vmatrix} a_{31} & a_{34} \\ a_{51} & a_{54} \end{vmatrix} = a_{31}a_{54} - a_{34,51} \quad \text{and} \quad (-1)^{3+5+1+4} \left| A_{3,5}^{1,4} \right| = -\left| A_{3,5}^{1,4} \right|.$$

157. When the row and column indices of the submatrix are the same, i.e., when the diagonal elements of the minor come from the diagonal of the matrix.

158. **(A)** Yes, since the diagonal elements belong to the diagonal of A.
(B) No, since a_{23} belongs to the diagonal of M_2 but not to A.
(C) Yes, since the diagonal elements belong to the diagonal of A.

159. The missing row indices are 1 and 3, and the missing column indices are 1 and 3. Hence

$$\begin{vmatrix} a_{11} & a_{13} \\ a_{31} & a_{33} \end{vmatrix}$$

is the complement of M_1. (In general, a minor is a principal minor iff its complement is a principal minor.)

160. $\begin{vmatrix} a_{32} & a_{34} \\ a_{42} & a_{44} \end{vmatrix}$ (not principal)

Chapter 6: Vector Spaces

161. The $+$ in $(a + b)\mathbf{u}$ denotes the addition of the two scalars a and b; hence it represents the addition operation in the field K. On the other hand, the $+$ in $a\mathbf{u} + b\mathbf{u}$ denotes the addition of the two vectors $a\mathbf{u}$ and $b\mathbf{u}$; hence it represents the operation of vector addition. Thus each $+$ represents a different operation.

162. Vector addition and scalar multiplication are defined by $(a_1, a_2, \ldots, a_n) + (b_1, b_2, \ldots, b_n) = (a_1 + b_1, a_2 + b_2, \ldots, a_n + b_n)$ and $k(a_1, a_2, \ldots, a_n) = (ka_1, ka_2, \ldots, ka_n)$ where a_i, b_i, $k \in K$. The zero vector in V is the n-tuple of zeros, $\mathbf{0} = (0, 0, \ldots, 0)$.

163. V is a vector space over K with respect to the operations of matrix addition and scalar multiplication.

164. V is a vector space over K with respect to the usual operations of addition of polynomials and multiplication by a constant.

165. (A) By axiom $[A_2]$ with $\mathbf{u} = 0$, we have $\mathbf{0} + \mathbf{0} = \mathbf{0}$. Hence by axiom $[M_1]$, $k\mathbf{0} = k(\mathbf{0} + \mathbf{0}) = k\mathbf{0} + k\mathbf{0}$. Adding $-k\mathbf{0}$ to both sides gives the desired result.
   (B) By a property of K, $0 + 0 = 0$. Hence by axiom $[M_2]$, $0\mathbf{u} = (0 + 0)\mathbf{u} = 0\mathbf{u} + 0\mathbf{u}$. Adding $-0\mathbf{u}$ to both sides yields the required result.
   (C) Suppose $k\mathbf{u} = \mathbf{0}$ and $k \neq 0$. Then there exists a scalar k^{-1} such that $k^{-1} k = 1$; hence $\mathbf{u} = 1\mathbf{u} = (k^{-1} k)\mathbf{u} = k^{-1} (k\mathbf{u}) = k^{-1} \mathbf{0} = \mathbf{0}$.
   (D) Using $\mathbf{u} + (-\mathbf{u}) = \mathbf{0}$, we obtain $\mathbf{0} = k\mathbf{0} = k[\mathbf{u} + (-\mathbf{u})] = k\mathbf{u} + k(-\mathbf{u})$. Adding $-k\mathbf{u}$ to both sides gives $-k\mathbf{u} = k(-\mathbf{u})$.
   Using $k + (-k) = 0$, we obtain $\mathbf{0} = 0\mathbf{u} = [k + (-k)]\mathbf{u} = k\mathbf{u} + (-k)\mathbf{u}$. Adding $-k\mathbf{u}$ to both sides yields $-k\mathbf{u} = (-k)\mathbf{u}$. Thus $(-k)\mathbf{u} = k(-\mathbf{u}) = -k\mathbf{u}$.

166. $\mathbf{0} = (0, 0, 0) \in W$ since $0 + 0 + 0 = 0$. Suppose $\mathbf{v} = (a, b, c)$, $\mathbf{w} = (a', b', c')$ belong to W, i.e., $a + b + c = 0$ and $a' + b' + c' = 0$. Then for any scalars k and k', $k\mathbf{v} + k'\mathbf{w} = k(a, b, c,) + k'(a', b', c') = (ka, kb, kc) + (k'a', k'b', k'c') = (ka + k'a', kb + k'b', kc + k'c')$ and, furthermore,

$$(ka + k'a') + (kb + k'b') + (kc + k'c') = k(a + b + c) + k'(a' + b' + c') = k0 + k'0 = 0.$$

Thus $k\mathbf{v} + k'\mathbf{w} \in W$, and so W is a subspace of R^3.

167. $0 \in W$ since all entries of 0 are 0 and hence equal. Now suppose $A = [a_{ij}]$ and $B = [b_{ij}]$ belong to W, i.e., $a_{ji} = a_{ij}$ and $b_{ji} = b_{ij}$. For any scalars $a, b \in K$, $aA + bB$ is the matrix whose ij-entry is $aa_{ij} + bb_{ij}$. But $aa_{ji} + bb_{ji} = aa_{ij} + bb_{ij}$. Thus $aA + bB$ is also symmetric, and so W is a subspace of V.

168. The system is equivalent to the matrix equation $A\mathbf{x} = \mathbf{0}$. Since $A\mathbf{0} = \mathbf{0}$, the zero vector $\mathbf{0} \in W$. Suppose $\mathbf{u}$ and $\mathbf{v}$ are vectors in W, i.e., $\mathbf{u}$ and $\mathbf{v}$ are solutions of the system. Then $A\mathbf{u} = \mathbf{0}$ and $A\mathbf{v} = \mathbf{0}$. Thus, for any scalars a and b in K, we have $A(a\mathbf{u} + b\mathbf{v}) = aA\mathbf{u} + bA\mathbf{v} = a\mathbf{0} + b\mathbf{0} = \mathbf{0} + \mathbf{0} = \mathbf{0}$. Hence $a\mathbf{u} + b\mathbf{v}$ is a solution of the system, i.e., $a\mathbf{u} + b\mathbf{v} \in W$. Thus W is a subspace of K^n.

169. We wish to express $\mathbf{v}$ as $\mathbf{v} = x\mathbf{e}_1 + y\mathbf{e}_2 + z\mathbf{e}_3$, with x, y, and z as yet unknown scalars. Thus we require

$$(1, -2, 5) = x(1, 1, 1) + y(1, 2, 3) + z(2, -1, 1)$$
$$= (x, x, x) + (y, 2y, 3y) + (2z, -z, z)$$
$$= (x + y + 2z, x + 2y - z, x + 3y + z).$$

Form the equivalent system of equations by setting corresponding components equal to each other, and then reduce to echelon form:

$$
\begin{array}{lll}
x + y + 2z = 1 & x + y + 2z = 1 & x + y + 2z = 1 \\
x + 2y - z = -2 \quad \text{or} & y - 3z = -3 \quad \text{or} & y - 3z = -3 \\
x + 3y + z = 5 & 2y - z = 4 & 5z = 10.
\end{array}
$$

Note that the above system is now triangular and so has a solution. Solve for the unknowns to obtain $x = -6$, $y = 3$, $z = 2$. Hence $\mathbf{v} = -6\mathbf{e}_1 + 3\mathbf{e}_2 + 2\mathbf{e}_3$.

170. Set $\mathbf{u} = x\mathbf{v} + y\mathbf{w}$: $(1, -2, k) = x(3, 0, -2) + y(2, -1, -5) = (3x + 2y, -y, -2x - 5y)$. Form the equivalent system of equations:

$$3x + 2y = 1, \qquad -y = -2, \qquad -2x - 5y = k.$$

By the first two equations, $x = -1$, $y = 2$. Substitute into the last equation to obtain $k = -8$.

171. Let $\mathbf{v} = (a, b, c)$ be an arbitrary vector in $\mathbf{R}^3$. Set $\mathbf{v} = x\mathbf{e}_1 + y\mathbf{e}_2 + z\mathbf{e}_3$ where x, y, z, are unknown scalars: $(a, b, c) = x(1, 0, 0) + y(0, 1, 0,) + z(0, 0, 1) = (x, y, z)$. Thus $x = a$, $y = b$, $z = c$. Hence $\mathbf{v}$ is a linear combination of $\mathbf{e}_1$, $\mathbf{e}_2$, $\mathbf{e}_3$; specifically, $\mathbf{v} = a\mathbf{e}_1 + b\mathbf{e}_2 + c\mathbf{e}_3$. Thus $\mathbf{e}_1$, $\mathbf{e}_2$, $\mathbf{e}_3$ span $\mathbf{R}^3$.

172. Matrices have the same row space if and only if their row canonical forms have the same nonzero rows; hence row reduce each matrix to row canonical form:

$$A = \begin{bmatrix} 1 & 1 & 5 \\ 2 & 3 & 13 \end{bmatrix} \sim \begin{bmatrix} 1 & 1 & 5 \\ 0 & 1 & 3 \end{bmatrix} \sim \begin{bmatrix} 1 & 0 & 2 \\ 0 & 1 & 3 \end{bmatrix}$$

$$B = \begin{bmatrix} 1 & -1 & -2 \\ 3 & -2 & -3 \end{bmatrix} \sim \begin{bmatrix} 1 & -1 & -2 \\ 0 & 1 & 3 \end{bmatrix} \sim \begin{bmatrix} 1 & 0 & 1 \\ 0 & 1 & 3 \end{bmatrix}$$

$$C = \begin{bmatrix} 1 & -1 & -1 \\ 4 & -3 & -1 \\ 3 & -1 & 3 \end{bmatrix} \sim \begin{bmatrix} 1 & -1 & -1 \\ 0 & 1 & 3 \\ 0 & 2 & 6 \end{bmatrix} \sim \begin{bmatrix} 1 & -1 & -1 \\ 0 & 1 & 3 \\ 0 & 0 & 0 \end{bmatrix} \sim \begin{bmatrix} 1 & 0 & 2 \\ 0 & 1 & 3 \\ 0 & 0 & 0 \end{bmatrix}.$$

Since the nonzero rows of the reduced form of A and of the reduced form of C are the same, A and C have the same row space. On the other hand, the nonzero rows of the reduced form of B are not the same as the others, and so B has a different row space.

173. Suppose $\mathbf{r} = (a_1, a_2, \ldots, a_m)$ and $B = [b_{ij}]$. Let $B_1, \ldots, B_m$ denote the rows of B and $B^1, \ldots, B^n$ its columns. Then

$$\mathbf{r}B = (\mathbf{r} \cdot B^1, \mathbf{r} \cdot B^2, \ldots, \mathbf{r} \cdot B^n)$$
$$= (a_1 b_{11} + a_2 b_{21} + \cdots + a_m b_{m1}, a_1 b_{12} + a_2 b_{22} + \cdots + a_m b_{m2}, \ldots, a_1 b_{1n} + a_2 b_{2n} + \cdots + a_m b_{mn})$$
$$= a_1(b_{11}, b_{12}, \ldots, b_{1n}) + a_2(b_{21}, b_{22}, \ldots, b_{2n}) + \cdots + a_m(b_{m1}, b_{m2}, \ldots, b_{mn})$$
$$= a_1 B_1 + a_2 B_2 + \cdots + a_m B_m.$$

Thus $\mathbf{r}B$ is a linear combination of the rows of B, as claimed.

174. Form the matrix A whose rows are the $\mathbf{u}_i$, and row reduce A to row canonical form:

$$A = \begin{bmatrix} 1 & 1 & -1 \\ 2 & 3 & -1 \\ 3 & 1 & -5 \end{bmatrix} \sim \begin{bmatrix} 1 & 1 & -1 \\ 0 & 1 & 1 \\ 0 & -2 & -2 \end{bmatrix} \sim \begin{bmatrix} 1 & 0 & -2 \\ 0 & 1 & 1 \\ 0 & 0 & 0 \end{bmatrix}.$$

Next form the matrix B whose rows are the $\mathbf{w}_i$, and row reduce B to row canonical form:

$$B = \begin{bmatrix} 1 & -1 & -3 \\ 3 & -2 & -8 \\ 2 & 1 & -3 \end{bmatrix} \sim \begin{bmatrix} 1 & -1 & -3 \\ 0 & 1 & 1 \\ 0 & 3 & 3 \end{bmatrix} \sim \begin{bmatrix} 1 & 0 & -2 \\ 0 & 1 & 1 \\ 0 & 0 & 0 \end{bmatrix}.$$

Since A and B have the same row canonical form, the row spaces of A and B are equal and so $U = W$.

175. Since U and W are subspaces, $\mathbf{0} \in U$ and $\mathbf{0} \in W$. Hence $\mathbf{0} = \mathbf{0} + \mathbf{0} \in U + W$. Suppose $\mathbf{v}, \mathbf{v}' \in U + W$. Then there exist $\mathbf{u}, \mathbf{u}' \in U$ and $\mathbf{w}, \mathbf{w}' \in W$ such that $\mathbf{v} = \mathbf{u} + \mathbf{w}$ and $\mathbf{v}' = \mathbf{u}' + \mathbf{w}'$. Since U and W are subspaces, $\mathbf{u} + \mathbf{u}' \in U$ and $\mathbf{w} + \mathbf{w}' \in W$ and, for any scalar k, $k\mathbf{u} \in U$ and $k\mathbf{w} \in W$. Accordingly, $\mathbf{v} + \mathbf{v}' = (\mathbf{u} + \mathbf{w}) + (\mathbf{u}' + \mathbf{w}') = (\mathbf{u} + \mathbf{u}') + (\mathbf{w} + \mathbf{w}') \in U + W$ and, for any scalar k, $k\mathbf{v} = k(\mathbf{u} + \mathbf{w}) = k\mathbf{u} + k\mathbf{w} \in U + W$. Thus $U + W$ is a subspace of V.

176. Any vector $(a, b, c) \in \mathbf{R}^3$ can be written as the sum of a vector in U and a vector in V in one and only one way:

$$(a, b, c) = (a, b, 0) + (0, 0, c).$$

Accordingly, $\mathbf{R}^3$ is the direct sum of U and W; that is, $\mathbf{R}^3 = U \oplus W$.

177. Note first that $U \cap W = \{0\}$, for $\mathbf{v} = (a, b, c) \in U \cap W$ implies that $a = b = c$ and $a = 0$, which implies $a = 0$, $b = 0$, $c = 0$, i.e., $\mathbf{v} = (0, 0, 0)$.
We also claim that $\mathbf{R}^3 = U + W$. For if $\mathbf{v} = (a, b, c) \in \mathbf{R}^3$, then $\mathbf{v} = (a, a, a) + (0, b - a, c - a)$ where $(a, a, a) \in U$ and $(0, b - a, c - a) \in W$. Both conditions, $U \cap W = \{0\}$ and $\mathbf{R}^3 = U + W$, imply $\mathbf{R}^3 = U \oplus W$.

178. Suppose $\mathbf{v}_1$ and $\mathbf{v}_2$ are dependent. Then there exist scalars a, b, not both 0, such that $a\mathbf{v}_1 + b\mathbf{v}_2 = \mathbf{0}$. Say $a \neq 0$. Then $\mathbf{v}_1 = (-b/a)\mathbf{v}_2$. Conversely, suppose $\mathbf{v}_1 = k\mathbf{v}_2$. Then $\mathbf{v}_1 - k\mathbf{v}_2 = \mathbf{0}$ where 1 is the coefficient of $\mathbf{v}_1$, and hence $\mathbf{v}_1$ and $\mathbf{v}_2$ are dependent.

179. Two vectors $\mathbf{u}$ and $\mathbf{v}$ in $\mathbf{R}^3$ are dependent if and only if they lie on the same line through the origin. Three vectors $\mathbf{u}$, $\mathbf{v}$, and $\mathbf{w}$ in $\mathbf{R}^3$ are dependent if and only if they lie on the same plane through the origin.

180. (A) No, since neither is a multiple of the other.
(B) Yes, for $\mathbf{u} = -2\mathbf{v}$.

181. Form the matrix whose rows are the given vectors, and reduce to echelon form using the elementary row operations:

$$\begin{bmatrix} 1 & -2 & 1 \\ 2 & 1 & -1 \\ 7 & -4 & 1 \end{bmatrix} \text{ to } \begin{bmatrix} 1 & -2 & 1 \\ 0 & 5 & -3 \\ 0 & 10 & -6 \end{bmatrix} \text{ to } \begin{bmatrix} 1 & -2 & 1 \\ 0 & 5 & -3 \\ 0 & 0 & 0 \end{bmatrix}.$$

Since the echelon matrix has a zero row, the vectors are dependent.

182. Form the matrix whose rows are given vectors and row reduce the matrix to echelon form:

$$
\begin{bmatrix} 1 & 2 & -3 \\ 1 & -3 & 2 \\ 2 & -1 & 5 \end{bmatrix}
\quad \text{to} \quad
\begin{bmatrix} 1 & 2 & -3 \\ 0 & -5 & 5 \\ 0 & -5 & 11 \end{bmatrix}
\quad \text{to} \quad
\begin{bmatrix} 1 & 2 & -3 \\ 0 & -5 & 5 \\ 0 & 0 & 6 \end{bmatrix}.
$$

Since the echelon matrix has no zero rows, the vectors are independent.

183. Suppose $x(\mathbf{u} + \mathbf{v}) + y(\mathbf{u} - \mathbf{v}) + z(\mathbf{u} - 2\mathbf{v} + \mathbf{w}) = \mathbf{0}$ where x, y, and z are scalars. Then $x\mathbf{u} + x\mathbf{v} + y\mathbf{u} - y\mathbf{v} + z\mathbf{u} - 2z\mathbf{v} + z\mathbf{w} = \mathbf{0}$, or $(x + y + z)\mathbf{u} + (x - y - 2z)\mathbf{v} + z\mathbf{w} = \mathbf{0}$. But $\mathbf{u}$, $\mathbf{v}$, and $\mathbf{w}$ are linearly independent; hence the coefficients in the above relation are each 0:

$$
\begin{aligned}
x + y + z &= 0 \\
x - y - 2z &= 0 \\
z &= 0.
\end{aligned}
$$

The only solution to the above system is $x = 0$, $y = 0$, $z = 0$. Hence $\mathbf{u} + \mathbf{v}$, $\mathbf{u} - \mathbf{v}$, and $\mathbf{u} - 2\mathbf{v} + \mathbf{w}$ are independent.

184. (A) If $\mathbf{u} = (a_1, a_2, \ldots, a_n)$, then $\mathbf{u} = a_1\mathbf{e}_1 + a_2\mathbf{e}_2 + \cdots + a_n\mathbf{e}_n$, Hence B spans V. The vectors in B form an echelon matrix, and hence are independent by Theorem 6.7. Thus B is a basis of $\mathbf{R}^n$.
(B) Since B has n vectors, $\dim(\mathbf{R}^n) = n$.

185. The three vectors form a basis if and only if they are independent. Thus form the matrix whose rows are the given vectors and row reduce to echelon form:

$$
\begin{bmatrix} 1 & 1 & 1 \\ 1 & 2 & 3 \\ 2 & -1 & 1 \end{bmatrix}
\quad \text{to} \quad
\begin{bmatrix} 1 & 1 & 1 \\ 0 & 1 & 2 \\ 0 & -3 & -1 \end{bmatrix}
\quad \text{to} \quad
\begin{bmatrix} 1 & 1 & 1 \\ 0 & 1 & 2 \\ 0 & 0 & 5 \end{bmatrix}.
$$

The echelon matrix has no zero rows; hence the three vectors are independent and so form a basis for $\mathbf{R}^3$.

186. Form the matrix whose rows are the given vectors and row reduce to echelon form:

$$
\begin{bmatrix} 1 & 1 & 2 \\ 1 & 2 & 5 \\ 5 & 3 & 4 \end{bmatrix}
\quad \text{to} \quad
\begin{bmatrix} 1 & 1 & 2 \\ 0 & 1 & 3 \\ 0 & -2 & -6 \end{bmatrix}
\quad \text{to} \quad
\begin{bmatrix} 1 & 1 & 2 \\ 0 & 1 & 3 \\ 0 & 0 & 0 \end{bmatrix}.
$$

The echelon matrix has a zero row (i.e., only two nonzero rows); hence the three vectors are dependent and so do not form a basis for $\mathbf{R}^3$.

187. We claim that $\{1, i\}$ is a basis of $\mathbf{C}$ over $\mathbf{R}$. For if $\mathbf{v} \in \mathbf{C}$, then $\mathbf{v} = a + bi = a \cdot 1 + b \cdot i$ where $a, b \in \mathbf{R}$; i.e., $\{1, i\}$ generates $\mathbf{C}$ over $\mathbf{R}$. Furthermore, if $x \cdot 1 + y \cdot i = 0$ or $x + yi = 0$, where $x, y \in \mathbf{R}$, then $x = 0$ and $y = 0$, i.e., $\{1, i\}$ is linearly independent over $\mathbf{R}$. Thus $\{1, i\}$ is a basis of $\mathbf{C}$ over $\mathbf{R}$, and so $\mathbf{C}$ is of dimension two over $\mathbf{R}$.

188. We claim that, for any n, $\{1, \pi, \pi^2, \ldots, \pi^n\}$ is linearly independent over $\mathbf{Q}$. For suppose $a_0 1 + a_1\pi + a_2\pi^2 + \cdots + a_n\pi^n = 0$, where the $a_i \in \mathbf{Q}$ and not all the a_i are 0. Then π is a root of the following nonzero polynomial over $\mathbf{Q}$: $a_0 + a_1 x + a_2 x^2 + \cdots + a_n x^n$. But it can be shown that π is a transcendental number, i.e., that π is not a root of any nonzero polynomial over $\mathbf{Q}$. Accordingly, the $n + 1$ real numbers $1, \pi, \pi^2, \ldots, \pi^2$ are linearly independent over $\mathbf{Q}$. Thus for any finite n, $\mathbf{R}$ cannot be of dimension n over $\mathbf{Q}$, i.e., $\mathbf{R}$ is of infinite dimension over $\mathbf{Q}$.

189. Since dim $(\mathbf{R}^3) = 3$, the dimension of W can only be 0, 1, 2, or 3. The following cases apply:
  (i) If dim $(W) = 0$, then $W = \{0\}$, a point.
  (ii) If dim $(W) = 1$, then W is a line through the origin.
  (iii) If dim $(W) = 2$, then W is a plane through the origin.
  (iv) If dim $(W) = 3$, then W is the entire space $\mathbf{R}^3$.

190. Note that $W \neq \mathbf{R}^3$ since, for example, $(1, 2, 3) \notin W$. Thus dim $(W) < 3$. Note that $\mathbf{u}_1 = (1, 0, -1)$ and $\mathbf{u}_2 = (0, 1, -1)$ are two independent vectors in W. Thus dim $(W) = 2$ and so $\mathbf{u}_1$ and $\mathbf{u}_2$ form a basis of W.

191. Reduce to echelon form the matrix whose rows are the given vectors:

$$\begin{bmatrix} 1 & 4 & -1 & 3 \\ 2 & 1 & -3 & -1 \\ 0 & 2 & 1 & -5 \end{bmatrix} \text{ to } \begin{bmatrix} 1 & 4 & -1 & 3 \\ 0 & -7 & -1 & -7 \\ 0 & 2 & 1 & -5 \end{bmatrix} \text{ to } \begin{bmatrix} 1 & 4 & -1 & 3 \\ 0 & -7 & -1 & -7 \\ 0 & 0 & 5 & -49 \end{bmatrix}.$$

The nonzero rows in the echelon matrix form a basis of W; hence dim $(W) = 3$. In particular, this means the original three vectors are linearly independent and also form a basis for W.

192. Reduce to echelon form the matrix whose rows are the given vectors:

$$\begin{bmatrix} 1 & -4 & -2 & 1 \\ 1 & -3 & -1 & 2 \\ 3 & -8 & -2 & 7 \end{bmatrix} \text{ to } \begin{bmatrix} 1 & -4 & -2 & 1 \\ 0 & 1 & 1 & 1 \\ 0 & 4 & 4 & 4 \end{bmatrix} \text{ to } \begin{bmatrix} 1 & -4 & -2 & 1 \\ 0 & 1 & 1 & 1 \\ 0 & 0 & 0 & 0 \end{bmatrix}.$$

The nonzero rows in the echelon matrix, $(1, -4, -2, 1)$ and $(0, 1, 1, 1)$, form a basis of W and so dim $(W) = 2$. In particular, this means that the original three vectors are linearly dependent.

193. Row reduce A to an echelon form:

$$A \text{ to } \begin{bmatrix} 1 & 2 & 0 & -1 \\ 0 & 2 & -3 & -1 \\ 0 & 4 & -6 & -2 \end{bmatrix} \text{ to } \begin{bmatrix} 1 & 2 & 0 & -1 \\ 0 & 2 & -3 & -1 \\ 0 & 0 & 0 & 0 \end{bmatrix}.$$

The nonzero rows of the echelon matrix form a basis of the row space of A, and hence the dimension of the row space of A is two. Thus rank$(A) = 2$.

194. Row reduce B to an echelon form:

$$B \text{ to } \begin{bmatrix} 1 & 3 & 1 & -2 & -3 \\ 0 & 1 & 2 & 1 & -1 \\ 0 & -3 & -6 & -3 & 3 \\ 0 & -1 & -2 & -1 & 1 \end{bmatrix} \text{ to } \begin{bmatrix} 1 & 3 & 1 & -2 & -3 \\ 0 & 1 & 2 & 1 & -1 \\ 0 & 0 & 0 & 0 & 0 \\ 0 & 0 & 0 & 0 & 0 \end{bmatrix}.$$

Since the echelon matrix has two nonzero rows, rank$(B) = 2$.

195. Since row rank equals column rank, we can form the transpose of A and then row reduce to echelon form:

$$\begin{bmatrix} 1 & 2 & -2 & -1 \\ 2 & 1 & -1 & 4 \\ -3 & 0 & 3 & -2 \end{bmatrix} \text{ to } \begin{bmatrix} 1 & 2 & -2 & -1 \\ 0 & -3 & 3 & 6 \\ 0 & 6 & -6 & -5 \end{bmatrix} \text{ to } \begin{bmatrix} 1 & 2 & -2 & -1 \\ 0 & -3 & 3 & 6 \\ 0 & 0 & 3 & 7 \end{bmatrix}.$$

Thus rank$(A) = 3$.

196. (A) The row space of AB is contained in the row space of B; hence rank$(AB) \leq$ rank (B).
(B) The column space of AB is contained in the column space of A; hence rank$(AB) \leq$ rank(A).

197. Reduce the system to echelon form:

$$\begin{aligned} x + 2y + 2z - s + 3t &= 0 \\ z + 2s - 2t &= 0 \\ 2z + 4s - 4t &= 0. \end{aligned} \quad \text{or} \quad \begin{aligned} x + 2y + 2z - s + 3t &= 0 \\ z + 2s - 2t &= 0. \end{aligned}$$

The system in echelon form has two (nonzero) equations in five unknowns, and hence the system has $5 - 2 = 3$ free variables, which are y, s, and t. Thus dim $(W) = 3$. To obtain a basis for W, set

(i) $y = 1, s = 0, t = 0$ to obtain the solution $\mathbf{v}_1 = (-2, 1, 0, 0, 0)$;
(ii) $y = 0, s = 1, t = 0$ to obtain the solution $\mathbf{v}_2 = (5, 0, -2, 1, 0)$;
(iii) $y = 0, s = 0, t = 1$ to obtain the solution $\mathbf{v}_3 = (-7, 0, 2, 0, 1)$.

The set $\{\mathbf{v}_1, \mathbf{v}_2, \mathbf{v}_3\}$ is a basis of the solution space W.

198. Reduce the system to echelon form:

$$\begin{aligned} x + 2y + z - 3t &= 0 \\ 2z + 5t &= 0 \\ 4z + 10t &= 0 \end{aligned} \quad \text{or} \quad \begin{aligned} x + 2y + z - 3t &= 0 \\ 2z + 5t &= 0. \end{aligned}$$

The free variables are y and t and dim $(W) = 2$. Set

(i) $y = 1$, $z = 0$ to obtain the solution $\mathbf{u}_1 = (-2, 1, 0, 0)$;

(ii) $y = 0$, $t = 2$ to obtain the solution $\mathbf{u}_2 = (11, 0, -5, 2)$.

Then $\{\mathbf{u}_1, \mathbf{u}_2\}$ is a basis of W. [We could have chosen $y = 0$, $t = 1$ in (ii), but such a choice would have introduced fractions into the solution.]

199. Reduce the system to echelon form:

$$
\begin{aligned}
x + 2y - 3z &= 0 \\
y + 7z &= 0 \quad \text{and then} \\
-3y + 5z &= 0
\end{aligned}
\qquad
\begin{aligned}
x + 2y - 3z &= 0 \\
y + 7z &= 0 \\
26z &= 0.
\end{aligned}
$$

The echelon system is in triangular form and hence has no free variables. Thus 0 is the only solution; that is, $W = \{0\}$. Accordingly, dim $(W) = 0$.

200. Let $\mathbf{v} = (x, y, z, t)$. Form the matrix M whose first rows are the given vectors and whose last row is $\mathbf{v}$ and then row reduce to echelon form:

$$
M = \begin{bmatrix}
1 & -2 & 0 & 3 \\
1 & -1 & -1 & 4 \\
1 & 0 & -2 & 5 \\
x & y & z & t
\end{bmatrix}
\quad \text{to} \quad
\begin{bmatrix}
1 & -2 & 0 & 3 \\
0 & -1 & -1 & 1 \\
0 & -2 & -2 & 2 \\
0 & 2x+y & z & -3x+t
\end{bmatrix}
\quad \text{to}
$$

$$
\begin{bmatrix}
1 & -2 & 0 & 3 \\
0 & 1 & -1 & 1 \\
0 & 0 & 2x+y+z & -5x-y+t \\
0 & 0 & 0 & 0
\end{bmatrix}.
$$

The original first three rows show that W has dimension two. Thus $\mathbf{v} \in W$ if and only if the additional row does not increase the dimension of the row space. Hence we set the last two entries in the third row on the right equal to 0 to obtain the required homogeneous system

$$
\begin{aligned}
2x + y + z &= 0 \\
5x + y - t &= 0.
\end{aligned}
$$

201. Since U and W are distinct, $U + W$ properly contains U and W; hence dim$(U + W) > 4$. But dim$(U + W)$ cannot be greater than 6, since dim $(V) = 6$. Hence we have two possibilities: (i) dim$(U + W) = 5$ or (ii) dim$(U + W) = 6$. By Theorem 6.11, dim$(U \cap W) =$ dim (U) + dim (W) − dim$(U + W) = 8 - $ dim$(U + W)$. Thus (i) dim$(U \cap W) = 3$ or (ii) dim$(U \cap W) = 2$.

202. Suppose $U = W$. Then $U \cap W = U = W$ and hence dim$(U \cap W) = 2$. Suppose $U \neq W$. Then $U + W$ properly contains U (and W). Hence dim$(U + W) > $ dim $(U) = 2$. But $U + W \subseteq \mathbf{R}^3$, which has dimension three. Therefore dim$(U + W) = 3$. Thus, by Theorem 6.11, dim$(U \cap W) = $ dim (U) + dim (W) − dim$(U + W) = 2 + 2 - 3 = 1$. That is, $U \cap W$ is a line through the origin.

Remark: The above agrees with the well-known result in solid geometry that the intersection of two distinct planes is a line.

203. (A) $U + W$ is the space spanned by all six vectors. Hence form the matrix whose rows are the given six vectors and then row reduce to echelon form:

$$
\begin{bmatrix}
1 & 1 & 0 & -1 \\
1 & 2 & 3 & 0 \\
2 & 3 & 3 & -1 \\
1 & 2 & 2 & -2 \\
2 & 3 & 2 & -3 \\
1 & 3 & 4 & -3
\end{bmatrix}
\text{ to }
\begin{bmatrix}
1 & 1 & 0 & -1 \\
0 & 1 & 3 & 1 \\
0 & 1 & 3 & 1 \\
0 & 1 & 2 & -1 \\
0 & 1 & 2 & -1 \\
0 & 2 & 4 & -2
\end{bmatrix}
\text{ or }
\begin{bmatrix}
1 & 1 & 0 & -1 \\
0 & 1 & 3 & 1 \\
0 & 1 & 2 & -1 \\
0 & 0 & 0 & 0 \\
0 & 0 & 0 & 0 \\
0 & 0 & 0 & 0
\end{bmatrix}
\text{ or }
\begin{bmatrix}
1 & 1 & 0 & -1 \\
0 & 1 & 3 & 1 \\
0 & 0 & -1 & -2 \\
0 & 0 & 0 & 0 \\
0 & 0 & 0 & 0 \\
0 & 0 & 0 & 0
\end{bmatrix}.
$$

The nonzero rows of the echelon matrix, $(1, 1, 0, -1)$, $(0, 1, 3, 1)$, and $(0, 0, -1, -2)$, form a basis of $U + W$, and so $\dim(U + W) = 3$.

(B) Reduce to echelon form the matrix whose rows span U:

$$
\begin{bmatrix}
1 & 1 & 0 & -1 \\
1 & 2 & 3 & 0 \\
2 & 3 & 3 & -1
\end{bmatrix}
\text{ to }
\begin{bmatrix}
1 & 1 & 0 & -1 \\
0 & 1 & 3 & 1 \\
0 & 1 & 3 & -1
\end{bmatrix}
\text{ to }
\begin{bmatrix}
1 & 1 & 0 & -1 \\
0 & 1 & 3 & 1 \\
0 & 0 & 0 & 0
\end{bmatrix}.
$$

The two nonzero rows of the echelon matrix, $(1, 1, 0, -1)$ and $(0, 1, 3, 1)$, form a basis of U and so $\dim(U) = 2$.

(C) Reduce to echelon form the matrix whose rows span W:

$$
\begin{bmatrix}
1 & 2 & 2 & -2 \\
2 & 3 & 2 & -3 \\
1 & 3 & 4 & -3
\end{bmatrix}
\text{ to }
\begin{bmatrix}
1 & 2 & 2 & -2 \\
0 & -1 & -2 & 1 \\
0 & 1 & 2 & -1
\end{bmatrix}
\text{ to }
\begin{bmatrix}
1 & 2 & 2 & -2 \\
0 & -1 & -2 & 1 \\
0 & 0 & 0 & 0
\end{bmatrix}.
$$

The two nonzero rows of the echelon matrix, $(1, 2, 2, -2)$ and $(0, -1, -2, 1)$, form a basis of W and so $\dim(W) = 2$.

(D) Use Theorem 6.11(ii): $\dim(U \cap W) = \dim(U) + \dim(W) - \dim(U + W) = 2 + 2 - 3 = 1$.

204. Since $V = U \oplus W$, we have $V = U + W$ and $U \cap W = \{0\}$. Thus $\dim(V) = \dim(U) + \dim(W) - \dim(U \cap W) = \dim(U) + \dim(W) - 0 = \dim(U) + \dim(W)$.

205. Since $U \not\subseteq W$, the intersection $U \cap W$ is properly contained in U and $U + W$ properly contains W. Thus $\dim(U \cap W) < \dim(U) = 1$. Hence $\dim(U \cap W) = 0$ and so $U \cap W = \{0\}$. Also, $\dim(U + W) > \dim(W) = 2$, but $U + W \subseteq (\mathbf{R}^3)$ where $\dim \mathbf{R}^3 = 3$. Thus $\dim(U + W) = 3$ and so $\mathbf{R}^3 = U + W$. The conditions $\mathbf{R}^3 = U + W$ and $U \cap W = \{0\}$ imply that $\mathbf{R}^3 = U \oplus W$.

206. **(A)** Set $\mathbf{v}$ as a linear combination of the f_i using the unknowns x, y, and z, i.e., set $\mathbf{v} = x\mathbf{f}_1 + y\mathbf{f}_2 + z\mathbf{f}_3$:

$$(3, 1, -4) = x(1, 1, 1) + y(0, 1, 1) + z(0, 0, 1)$$
$$= (x, x, x) + (0, y, y) + (0, 0, z)$$
$$= (x, x + y, x + y + z).$$

Then set the corresponding components equal to each other to obtain the equivalent system of equations

$$
\begin{aligned}
x &= 3 \\
x + y &= 1 \\
x + y + z &= -4
\end{aligned}
$$

the solution is $x = 3$, $y = -2$, $z = -5$. Thus $[\mathbf{v}]_f = [3, -2, -5]$.

(B) Set $\mathbf{v} = x\mathbf{e}_1 + y\mathbf{e}_2 + z\mathbf{e}_3$ using unknowns x, y, z:

$$(3, 1, -4) = x(1, 0, 0) + y(0, 1, 0) + z(0, 0, 1) = (x, y, z).$$

Set corresponding components equal to each other to obtain $x = 3$, $y = 1$, $z = -4$, Thus $[\mathbf{v}]_e = [3, 1, -4]$. In other words, relative to the usual basis, $[\mathbf{v}]_e$ has the same components as $\mathbf{v}$.

207. **(A)** Set $\mathbf{v} = x\mathbf{u}_1 + y\mathbf{u}_2$ to obtain $(2, 3) = x(2, 1) + y(1, -1) = (2x + y, x - y)$. Set corresponding components equal to each other to obtain the equations $2x + y = 2$ and $x - y = 3$. Solve to obtain $x = \frac{5}{3}$, $y = -\frac{4}{3}$. Thus $[\mathbf{v}] = [\frac{5}{3}, -\frac{4}{3}]$.

(B) Set $\mathbf{u} = x\mathbf{u}_1 + y\mathbf{u}_2$ to obtain $(4, -1) = (2x + y, x - y)$. Solve $2x + y = 4$ and $x - y = -1$ to get $x = 1$, $y = 2$. Hence $[\mathbf{u}] = [1, 2]$.

(C) Set $\mathbf{w} = x\mathbf{u}_1 + y\mathbf{u}_2$ to obtain $(3, -3) = (2x + y, x - y)$. Solve $2x + y = 3$ and $x - y = -3$ to get $x = 0$, $y = 3$. Thus $[\mathbf{w}] = [0, 3]$.

(D) Set $\mathbf{v} = x\mathbf{u}_1 + y\mathbf{u}_2$ to obtain $(a, b) = (2x + y, x - y)$. Solve $2x + y = a$ and $x - y = b$ to obtain $x = (a + b)/3$, $y = (a - 2b)/3$. Thus $[\mathbf{v}] = [(a + b)/3, (a - 2b)/3]$.

208. Write each new basis vector from S' as a linear combination of the original basis vectors in S:

$$\begin{bmatrix} 1 \\ -1 \end{bmatrix} = x\begin{bmatrix} 1 \\ 2 \end{bmatrix} + y\begin{bmatrix} 3 \\ 5 \end{bmatrix} \quad \text{or} \quad \begin{aligned} x + 3y &= 1 \\ 2x + 5y &= -1 \end{aligned} \quad \text{yielding} \quad x = -8, y = 3$$

$$\begin{bmatrix} 1 \\ -2 \end{bmatrix} = x\begin{bmatrix} 1 \\ 2 \end{bmatrix} + y\begin{bmatrix} 3 \\ 5 \end{bmatrix} \quad \text{or} \quad \begin{aligned} x + 3y &= 1 \\ 2x + 5y &= -2 \end{aligned} \quad \text{yielding} \quad x = -11, y = 4.$$

Thus

$$\begin{aligned} \mathbf{v}_1 &= -8\mathbf{u}_1 + 3\mathbf{u}_2 \\ \mathbf{v}_2 &= -11\mathbf{u}_1 + 4\mathbf{u}_2 \end{aligned} \quad \text{and hence} \quad P = \begin{bmatrix} -8 & -11 \\ 3 & 4 \end{bmatrix}.$$

Note that the coordinates of $\mathbf{v}_1$ and $\mathbf{v}_2$ are columns, not rows, of the change-of-basis matrix P.

209. Here we write each of the "old" basis vectors $\mathbf{u}_1$ and $\mathbf{u}_2$ of S' as a linear combination of the "new" basis vectors $\mathbf{v}_1$ and $\mathbf{v}_2$ of S'. This yields

$$\mathbf{u}_1 = 4\mathbf{v}_1 - 3\mathbf{v}_2 \quad \text{and hence} \quad Q = \begin{bmatrix} 4 & 11 \\ -3 & -8 \end{bmatrix}.$$
$$\mathbf{u}_2 = 11\mathbf{v}_1 - 8\mathbf{v}_2$$

As expected, $Q = P^{-1}$. (In fact, we could have obtained Q by simply finding P^{-1}.)

210. (A) Let $(2, 1) = x(1, 2) + y(3, 5)$. Hence $x + 3y = 2$, $2x + 5y = 1$. This yields $x = -7, y = 3$.
Thus $[\mathbf{v}]_S = [-7, 3]^T$.
　　(B) Let $(2, 1) = x(1, -1) + y(1, -2)$. Hence $x + y = 2$, $-x - 2y = 1$. This yields $x = 5$, $y = -3$.
Thus $[\mathbf{v}]_S' = [5, -3]^T$.
　　(C) $\begin{bmatrix} 4 & 11 \\ -3 & -8 \end{bmatrix}\begin{bmatrix} -7 \\ 3 \end{bmatrix} = \begin{bmatrix} 5 \\ -3 \end{bmatrix}.$

Chapter 7: Linear Mappings

211. (A) Substitute $k = 0$ into $F(k\mathbf{v}) = kF(\mathbf{v})$ to get $F(\mathbf{0}) = \mathbf{0}$.
　　(B) Using $F(k\mathbf{u}) = kF(\mathbf{u})$, we have $F(-\mathbf{u}) = F[(-1)\mathbf{u}] = (-1)F(\mathbf{u}) = -F(\mathbf{u})$.

212. Suppose F is linear; then $F(a\mathbf{v} + b\mathbf{w}) = F(a\mathbf{v}) + F(b\mathbf{w}) = aF(\mathbf{v}) = + bF(\mathbf{w})$. Conversely, suppose (i) holds. For $a = 1$ and $b = 1$ we get $F(\mathbf{v} + \mathbf{w}) = F(\mathbf{v}) + F(\mathbf{w})$, and for $b = 0$ we get $F(a\mathbf{v}) = aF(\mathbf{v})$; hence F is linear.

213. By properties of matrices, $T(\mathbf{v} + \mathbf{w}) = A(\mathbf{v} + \mathbf{w}) = A\mathbf{v} + A\mathbf{w} = T(\mathbf{v}) + T(\mathbf{w})$ and $T(k\mathbf{v}) = A(k\mathbf{v}) = kA\mathbf{v} = kT(\mathbf{v})$, where $\mathbf{v}, \mathbf{w} \in K^n$ and $k \in K$. Thus T is linear.
Remark: The above type of linear mapping will occur again and again. In fact, in the next chapter we show that every linear mapping from one finite-dimensional vector space into another can be represented as a linear mapping of the above type.

214. For any vectors $\mathbf{v}, \mathbf{w} \in V$ and any scalar $a, b \in K$, $(G \circ F)(a\mathbf{v} + b\mathbf{w}) = G(F(a\mathbf{v} + b\mathbf{w})) = G(aF(\mathbf{v}) + bF(\mathbf{w})) = aG(F(\mathbf{v})) = bG(F(\mathbf{w})) = a(G \circ F)(\mathbf{v}) + b(G \circ F)(\mathbf{w})$. Thus $G \circ F$ is linear.

215. Suppose $\mathbf{u}, \mathbf{u}' \in U$. Since F is one-to-one and onto, there exist unique vectors $\mathbf{v}, \mathbf{v}' \in V$ for which $F(\mathbf{v}) = \mathbf{u}$ and $F(\mathbf{v}') = \mathbf{u}'$. Since F is linear, we also have $F(\mathbf{v} + \mathbf{v}') = F(\mathbf{v}) + F(\mathbf{v}') = \mathbf{u} + \mathbf{u}'$ and $F(k\mathbf{v}) = kF(\mathbf{v}) = k\mathbf{u}$. By definition of the inverse mapping, $F^{-1}(\mathbf{u}) = \mathbf{v}$, $F^{-1}(\mathbf{u}') = \mathbf{v}'$, $F^{-1}(\mathbf{u} + \mathbf{u}') = \mathbf{v} + \mathbf{v}'$, and $F^{-1}(k\mathbf{u}) = k\mathbf{v}$. Then $F^{-1}(\mathbf{u} + \mathbf{u}') = \mathbf{v} + \mathbf{v}' = F^{-1}(\mathbf{u}) + F^{-1}(\mathbf{u}')$ and $F^{-1} = (k\mathbf{u}) = k\mathbf{v} = kF^{-1}(\mathbf{u})$ and thus F^{-1} is linear.

216. (A) Since $(3, 1)$ and $(1, 1)$ are linearly independent, they form a basis for $\mathbf{R}^2$; hence such a linear map T exists and is unique.

(B) First write (a, b) as a linear combination of $(3, 1)$ and $(1, 1)$, using unknown scalars x and y:

$$(a, b) = x(3, 1) + y(1, 1).$$

Hence

$$(a, b) = (3x, x) + (y, y) = (3x + y, x + y) \qquad \text{and so} \qquad \begin{cases} 3x + y = a \\ x + y = b. \end{cases}$$

Solving for x and y in terms of a and b yields $x = \frac{1}{2}a - \frac{1}{2}b$ and $y = -\frac{1}{2}a + \frac{3}{2}b$. Therefore $T(a, b) = xT(3, 1) + yT(1, 1) = x(2, -4) + y(0, 2) = (2x, -4x) + (0, 2y) = (2x, -4x + 2y) = (a - b, 5b - 3a)$.

(C) Use the formula for T to get $T(7, 4) = (7 - 4, 20 - 21) = (3, -1)$.

(D) Set $T(a, b) = (5, -3)$ and solve for a and b. We get $(a - b, -3a + 5b) = (5, -3)$, so $a - b = 5, -3a + 5b = -3$. Then $a = 11, b = 6$. Thus $F^{-1}(5, -3) = (11, 6)$.

217. (A) Since $F(\mathbf{0}) = \mathbf{0}, \mathbf{0} \in \text{Ker}(F)$. Now suppose $\mathbf{v}, \mathbf{w} \in \text{Ker}(F)$ and $a, b \in K$. Since $\mathbf{v}$ and $\mathbf{w}$ belong to the kernel of F, $F(\mathbf{v}) = \mathbf{0}$ and $F(\mathbf{w}) = \mathbf{0}$. Thus $F(a\mathbf{v} + b\mathbf{w}) = aF(\mathbf{v}) + bF(\mathbf{w}) = a\mathbf{0} + b\mathbf{0} = \mathbf{0}$ and so $a\mathbf{v} + b\mathbf{w} \in \text{Ker}(F)$. Thus the kernel of F is a subspace of V.

(B) Since $F(\mathbf{0}) = \mathbf{0}, \mathbf{0} \in \text{Im}(F)$. Now suppose $\mathbf{u}, \mathbf{u}' \in \text{Im}(F)$ and $a, b \in K$. Since $\mathbf{u}$ and $\mathbf{u}'$ belong to the image of F, there exist vectors $\mathbf{v}, \mathbf{v}' \in V$ such that $F(\mathbf{v}) = \mathbf{u}$ and $F(\mathbf{v}') = \mathbf{u}'$. Then $F(a\mathbf{v} + b\mathbf{v}') = aF(\mathbf{v}) + bF(\mathbf{v}') = a\mathbf{u} + b\mathbf{u}' \in \text{Im}(F)$. Thus the image of F is a subspace of U.

218. Suppose $\mathbf{u} \in \text{Im} F$; then $F(\mathbf{v}) = \mathbf{u}$ for some vector $\mathbf{v} \in V$. Since $\mathbf{v}_1, \ldots, \mathbf{v}_n$ span V and since $\mathbf{v} \in V$, there exist scalars $a_1, \ldots, a_n$ for which $\mathbf{v} = a_1\mathbf{v}_1 + a_2\mathbf{v}_2 + \cdots + a_n\mathbf{v}_n$. Accordingly,

$$\mathbf{u} = F(\mathbf{v}) = F(a_1\mathbf{v}_1 + a_2\mathbf{v}_2 + \cdots + a_n\mathbf{v}_n) = a_1 F(\mathbf{v}_1) + a_2 F(\mathbf{v}_2) + \cdots + a_n F(\mathbf{v}_n).$$

Thus the vectors $F(\mathbf{v}_1), \ldots, F(\mathbf{v}_n)$ span $\text{Im}(F)$.

219. (A) Find the image of vectors that span the domain $\mathbf{R}^3$:

$$T(1, 0, 0) = (1, 0, 1), \qquad T(0, 1, 0) = (2, 1, 1), \qquad T(0, 0, 1) = (-1, 1, -2).$$

The images span the image U of T; hence form the matrix whose rows are the image vectors and row reduce to echelon form:

$$\begin{bmatrix} 1 & 0 & 1 \\ 2 & 1 & 1 \\ -1 & 1 & -2 \end{bmatrix} \quad \text{to} \quad \begin{bmatrix} 1 & 0 & 1 \\ 0 & 1 & -1 \\ 0 & 1 & -1 \end{bmatrix} \quad \text{to} \quad \begin{bmatrix} 1 & 0 & 1 \\ 0 & 1 & -1 \\ 0 & 0 & 0 \end{bmatrix}.$$

Thus $\{(1, 0, 1), (0, 1, -1)\}$ is a basis of U, and so $\dim(U) = 2$.

(B) Set $T(\mathbf{v}) = 0$ where $\mathbf{v} = (x, y, z)$: $T(x, y, z) = (x + 2y - z, y + z, x + y - 2z) = (0, 0, 0)$. Set corresponding components equal to each other to form the homogeneous system whose solution space is the kernel W of T:

$$\begin{array}{l} x + 2y - z = 0 \\ y + z = 0 \\ x + y - 2z = 0 \end{array} \quad \text{or} \quad \begin{array}{l} x + 2y - z = 0 \\ y + z = 0 \\ -y - z = 0 \end{array} \quad \text{or} \quad \begin{array}{l} x + 2y - z = 0 \\ y + z = 0. \end{array}$$

227. Set $H(x, y, z) = (a, b, c)$ and then solve for x, y, z in terms of a, b, c:

$$x + y - 2z = a \qquad\qquad x + y - 2z = a$$
$$x + 2y + z = b \qquad \text{or} \qquad y + 3z = b - a$$
$$2x + 2y - 3z = c \qquad\qquad z = c - 2a.$$

Solving for x, y, z yields $x = -8a - b + 5c$, $y = 5a + b - 3c$, $z = -2a + c$. Thus $H^{-1}(a, b, c) = (-8a - b + 5c, 5a + b - 3c, -2a + c)$ or, replacing a, b, c by x, y, z, respectively, $H^{-1}(x, y, z) = (-8x - y + 5z, 5x + y - 3z, -2x + z)$.

228. (A) For any vectors $\mathbf{v}, \mathbf{w} \in V$ and any scalars $a, b \in K$,

$$(F + G)(a\mathbf{v} + b\mathbf{w}) = F(a\mathbf{v} + b\mathbf{w}) + G(a\mathbf{v} + b\mathbf{w})$$
$$= aF(\mathbf{v}) + bF(\mathbf{w}) + aG(\mathbf{v}) + bG(\mathbf{w})$$
$$= a(F(\mathbf{v}) + G(\mathbf{v})) + b(F(\mathbf{w}) + G(\mathbf{w}))$$
$$= a(F + G)(\mathbf{v}) + b(F + G)(\mathbf{w}).$$

Thus $F + G$ is linear.

(B) For any vectors $\mathbf{v}, \mathbf{w} \in V$ and any scalars $a, b \in K$,

$$(kF)(a\mathbf{v} + b\mathbf{w}) = kF(a\mathbf{v} + b\mathbf{w}) = k(aF(\mathbf{v}) + bF(\mathbf{w})) = akF(\mathbf{v}) + bkF(\mathbf{w}) = a(kF)(\mathbf{v}) + b(kF)(\mathbf{w}).$$

Thus kF is linear.

229. (A) $(F + G)(\mathbf{v}) = F(\mathbf{v}) + G(\mathbf{v}) = F(2, 3, 4) + G(2, 3, 4) = (4, 7) + (-2, 3) = (2, 10)$
(B) $(3F)(\mathbf{v}) = 3F(\mathbf{v}) = 3F(2, 3, 4) = 3(4, 7) = (12, 21)$

230. (A) $(F + G)(x, y, z) = F(x, y, z) + G(x, y, z) = (2x, y + z) + (x - z, y) = (3x - z, 2y + z)$
(B) $(3F)(x, y, z) = 3F(x, y, z) = 3(2x, y + z) = (6x, 3y + 3z)$
(C) $(2F - 5G)(x, y, z) = 2F(x, y, z) - 5G(x, y, z) = 2(2x, y + z) - 5(x - z, y) = (4x, 2y + 2z) + (-5x + 5z, -5y) = (-x + 5x, -3y + 2z)$

231. (A) $(H \circ F)(\mathbf{v}) = H(F(\mathbf{v})) = H(F(4, -1, 5)) = H(-1, 9) = (9, -2)$
(B) $(H \circ G)(\mathbf{w}) = H(G(\mathbf{w})) = H(G(3, 4, 1)) = H(2, 1) = (-1, 4)$

232. (A) $(F + G)(x, y, z) = F(x, y, z) + G(x, y, z) = (y, x + z) + (2z, x - y) = (y + 2z, 2x - y + z)$
(B) $(H \circ F)(x, y, z) = H(F(x, y, z)) = H(y, x + z) = (x + z, 2y)$
(C) $(H \circ G)(x, y, z) = H(G(x, y, z)) = H(2z, x - y) = (x - y, 4z)$
(D) Using (a), $H \circ (F + G)(x, y, z) = H((F + G)(x, y, z)) = H(y + 2z, 2x - y + z) = (2x - y + z, 2y + 4z)$
(E) $H^2(x, y) = H(H(x, y)) = H(y, 2x) = (2x, 2y)$

233. (A) $(F + G)(x, y, z) = F(x, y, z) + G(x, y, z) = (x + y + z, x + y) + (2x + z, x + y) = (3x + y + 2z, 2x + 2y)$
(B) $(F + H)(x, y, z) = F(x, y, z) + H(x, y, z) = (x + y + z, x + y) + (2y, x) = (x + 3y + z, 2x + y)$

(C) $G \circ F$ is not defined, since the co-domain of F is not the domain of G.

(D) $(3G + 2H)(x, y, z) = 3G(x, y, z) + 2H(x, y, z) = 3(2x + z, x + y) + 2(2y, x) = (6x + 4y + 3z, 5x + 3y)$

234. Suppose, for scalars $a, b, c \in K$,

$$aF + bG + cH = 0. \tag{1}$$

(Here **0** is the zero mapping.) For $\mathbf{e}_1 = (1, 0, 0) \in \mathbf{R}^3$, we have $(aF + bG + cH)(\mathbf{e}_1) = aF(1, 0, 0) + bG(1, 0, 0) + cH(1, 0, 0) = a(1, 1) + b(2, 1) + c(0, 1) = (a + 2b, a + b + c)$ and $\mathbf{0}(\mathbf{e}_1) = (0, 0)$. Thus by (1), $(a + 2b, a + b + c) = (0, 0)$ and so

$$a + 2b = 0 \quad \text{and} \quad a + b + c = 0. \tag{2}$$

Similarly, for $\mathbf{e}_2 = (0, 1, 0) \in \mathbf{R}^2$, we have $(aF + bG + cH)(\mathbf{e}_2) = aF(0, 1, 0) + bG(0, 1, 0) + cH(0, 1, 0) = a(1, 1) + b(0, 1) + c(2, 0) = (a + 2c, a + b) = \mathbf{0}(\mathbf{e}_2) = (0, 0)$. Thus

$$a + 2c = 0 \quad \text{and} \quad a + b = 0. \tag{3}$$

Using (2) and (3), we obtain

$$a = 0, \quad b = 0, \quad c = 0. \tag{4}$$

Since (1) implies (4), the mappings F, G, and H are linearly independent.

235. Since $\dim(\mathbf{R}^3) = 3$ and $\dim(\mathbf{R}^2) = 2$, by Theorem 7.4, $\dim(\text{Hom}(\mathbf{R}^3, \mathbf{R}^2)) = 3 \cdot 2 = 6$.

236. Suppose $\dim (V) = n$. Then $\dim(A(V)) = n^2$. Thus only the square integers can be the dimension of $A(V)$, that is, 9, 25, 36, 64, and 100.

237. (A) Since composition is the multiplication operation in $A(V)$, we have $T^2 = T \circ T$, $T^3 = T \circ T \circ T$, ….

(B) The operator $p(T)$ is defined by $p(T) = a_0 I + a_1 T + a_2 T^2 + \cdots + a_n T^n$. (For a scalar $k \in K$, the operator kI is frequently denoted simply by k.) In particular, if $p(T) = 0$, the zero mapping, then T is said to be *a zero* of the polynomial $p(x)$.

238. (A) $(5S - 3T)(x, y) = 5S(x, y) - 3T(x, y) = 5(x + y, 0) - 3(-y, x) = (5x + 5y, 0) + (3y, -3x) = (5x + 8y, -3x)$

(B) $(ST)(x, y) = S(T(x, y)) = S(-y, x) = (x - y, 0)$

(C) $(TS)(x, y) = T(S(x, y)) = T(x + y, 0) = (0, x + y)$

239. (A) $S^2(x, y) = S(S(x, y)) = S(x + y, 0) = (x + y, 0) = S(x, y)$. Thus $S^2 = S$.

(B) $T^2(x, y) = T(T(x, y)) = T(-y, x) = (-x, -y) = -(x, y) = -I(x, y)$. Thus $T^2 = -I$.

240. (A) $T^2(x, y) = T(T(x, y)) = T(x + 2y, 3x + 4y)$
$$= [(x + 2y) + 2(3x + 4y), 3(x + 2y) + 4(3x + 4y)] = (7x + 10y, 15x + 22y)$$

(B) $T^3(x, y) = T(T^2(x, y)) = T(7x + 10y, 15x + 22y)$
$$= [(7x + 10y) + 2(15x + 22y), 3(7x + 10y) + 4(15x + 22y)]$$
$$= (37x + 54y, 81x + 118y)$$

241. (A) By definition, $f(T) = T^2 - 3T + 4I$. Thus

$$f(T)(x, y) = T^2 - 3T + 4I)(x, y) = T^2(x, y) - 3T(x, y) + 4I(x, y)$$
$$= (7x + 10y, 15x + 22y) + (-3x - 6y, -9x - 12y) + (4x, 4y)$$
$$= (8x + 4y, 6x + 14y).$$

(B) No, since $f(T)$ is not the zero map.

242. (A) $g(T)(x, y) = (T^2 - 5T - 2I)(x, y) = T^2(x, y) - 5T(x, y) - 2I(x, y) = (7x + 10y, 15x + 22y) + (-5x - 10y, -15x - 20y) + (-2x, -2y) = (0, 0)$
(B) Yes, since $g(T) = 0$, the zero map.

243. We first show that $V = U + W$. Let $\mathbf{v} \in V$. Set $\mathbf{u} = E(\mathbf{v})$ and $\mathbf{w} = \mathbf{v} - E(\mathbf{v})$. Then $\mathbf{v} = E(\mathbf{v}) + \mathbf{v} - E(\mathbf{v}) = \mathbf{u} + \mathbf{w}$. By definition, $\mathbf{u} = E(\mathbf{v}) \in U$, the image of E. We now show that $\mathbf{w} \in W$, the kernel of E:

$$E(\mathbf{w}) = E(\mathbf{v} - E(\mathbf{v})) = E(\mathbf{v}) - E^2(\mathbf{v}) = E(\mathbf{v}) - E(\mathbf{v}) = \mathbf{0}$$

and thus $\mathbf{w} \in W$. Hence $V = U + W$.
   We next show that $U \cap W = \{0\}$. Let $\mathbf{v} \in U \cap W$. Since $\mathbf{v} \in U$, $E(\mathbf{v}) = \mathbf{v}$. Since $\mathbf{v} \in W$, $E(\mathbf{v}) = \mathbf{0}$. Thus $\mathbf{v} = E(\mathbf{v}) = \mathbf{0}$ and so $U \cap W = \{0\}$.
The above two properties imply that $V = U \oplus W$.

244. Let V be the vector space of polynomials over K, and let T be the operator on V defined by $T(a_0 + a_1 t + \cdots + a_n t^n) = a_0 t + a_1 t^2 + \cdots + a_n t^{n+1}$; that is, T increases the exponent of t in each term by 1. Now T is a linear mapping and is nonsingular. However, T is not onto and so is not invertible.

245. (A) We need only show that T is nonsingular, i.e., that $T^{-1}(0, 0) = (0, 0)$. Set $T(x, y) = (0, 0)$ and solve for x and y. We have $T(x, y) = (y, 2x - y) = (0, 0)$, or $2x - y = 0$, or $y = 0$. The only solution is $x = 0$, $y = 0$. Thus Ker $(T) = \{0\}$ and so T is nonsingular; hence T is invertible.
(B) Set $T(x, y) = (s, t)$ [and hence $T^{-1}(s, t) = (x, y)$]. We have $T(x, y) = (y, 2x - y) = (s, t)$ and so $y = s$, $2x - y = t$. Solving for x and y in terms of s and t, we obtain $x = \frac{1}{2}s + \frac{1}{2}t$, $y = s$. Thus T^{-1} is given by the formula $T^{-1}(s, t) = (\frac{1}{2}s + \frac{1}{2}t, s)$.
(C) Use the formula for T to get $T(6, 2) = (2, 12 - 2) = (2, 10)$.
(D) Use the formula for T^{-1} to get $T^{-1}(6, 2) = (3 + 1, 6) = (4, 6)$.

246. (A) Let $W = $ Ker (T). We need only show that T is nonsingular, i.e., that $W = \{0\}$. Set $T(x, y, z) = (0, 0, 0)$; that is, $T(x, y, z) = (2x, 4x - y, 2x + 3y - z) = (0, 0, 0)$. Thus W is the solution space of the homogeneous system $2x = 0$, $4x - y = 0$, $2x + 3y - z = 0$, which has only the trivial solution $(0, 0, 0)$. Thus $W = \{0\}$; hence T is nonsingular and so T is invertible.
(B) Set $T(x, y, z) = (r, s, t)$ [and so $T^{-1}(r, s, t) = (x, y, z)$]. We have $(2x, 4x - y, 2x + 3y - z) = (r, s, t)$, or $2x = r$, $4x - y = s$, $2x + 3y - z = t$. Solve for x, y, z in terms of r, s, t to get $x = \frac{1}{2}r$, $y = 2r - s$, $z = 7r - 3s - t$. Thus $T^{-1}(r, s, t) = (\frac{1}{2}r, 2r - s, 7r - 3s - t)$.
(C) Use the formula for T^{-1} to get $T^{-1}(2, 4, 6) = (1, 4 - 4, 14 - 12 - 6) = (1, 0, -4)$.

247. (A) Set $T(x, y, z) = (0, 0, 0)$ to get the homogeneous system $x + z = 0, x - z = 0, y = 0$, whose only solution is $x = 0, y = 0, z = 0$. Thus T is nonsingular and so T is invertible.

(b) Set $T(x, y, z) = (r, s, t)$ to get $x + z = r, x - z = s, y = t$. Solve for x, y, z to get $x = \frac{1}{2}r + \frac{1}{2}s, y = t, z = \frac{1}{2}r - \frac{1}{2}s$. Thus $T^{-1}(r, s, t) = (\frac{1}{2}r + \frac{1}{2}s, t, \frac{1}{2}r - \frac{1}{2}s)$, or $T^{-1}(x, y, z) = (\frac{1}{2}x + \frac{1}{2}y, z, \frac{1}{2}x - \frac{1}{2}y)$.

(C) Use the formula for T^{-1} to get $T^{-1}(2, 4, 6) = (1 + 2, 6, 1 - 2) = (3, 6, -1)$.

248. The identity operator $I \in A(V)$ is invertible and $S = I^{-1}SI$. Hence $S \sim S$.

249. Suppose $S \sim T$; say, $S = P^{-1}TP$ where P is invertible. Then P^{-1} is invertible and $T = PSP^{-1} = (P^{-1})^{-1}SP^{-1}$. Hence $T \sim S$.

250. Suppose $F \sim G$ and $G \sim H$; say, $F = P^{-1}GP$ and $G = Q^{-1}HQ$ where P and Q are invertible. Then QP is invertible and $F = P^{-1}GP = P^{-1}(Q^{-1}HQ)P = (P^{-1}Q^{-1})H(QP) = (QP)^{-1}H(QP)$. Hence $F \sim H$.

Chapter 8: Matrices and Linear Mappings

251. (A) $F(\mathbf{u}_1) = F(2, 1) = (4 - 5, 6 + 1) = (-1, 7)$.

(B) $\begin{bmatrix} -1 \\ 7 \end{bmatrix} = x\begin{bmatrix} 2 \\ 1 \end{bmatrix} + y\begin{bmatrix} 3 \\ 2 \end{bmatrix}$ or $\begin{aligned} 2x + 3y &= -1 \\ x + 2y &= 7. \end{aligned}$

The solution is $x = -23, y = 15$. Thus $F(\mathbf{u}_1) = -23\mathbf{u}_1 + 15\mathbf{u}_2$.

252. (A) $F(\mathbf{u}_2) = F(3, 2) = (6 - 10, 9 + 2) = (-4, 11)$.

(B) $\begin{bmatrix} -4 \\ 11 \end{bmatrix} = x\begin{bmatrix} 2 \\ 1 \end{bmatrix} + y\begin{bmatrix} 3 \\ 2 \end{bmatrix}$ or $\begin{aligned} 2x + 3y &= -4 \\ x + 2y &= 7. \end{aligned}$

The solution is $x = -29, y = 18$. Thus $F(\mathbf{u}_2) = -29\mathbf{u}_1 + 18\mathbf{u}_2$.

253. Write the coordinates of $F(\mathbf{u}_1)$ and $F(\mathbf{u}_2)$ as columns to get

$$[F] = \begin{bmatrix} -23 & -29 \\ 15 & 18 \end{bmatrix}.$$

Remark: It is usually advantageous to first find the coordinates of an arbitrary vector $(a, b) \in \mathbf{R}^2$ in the given basis, i.e., to first solve $(a, b) = x\mathbf{u}_1 + y\mathbf{u}_2 = x(2, 1) + y(3, 2)$ to obtain x and y in terms of a and b. This will be done in subsequent problems.

254. We have

$$\begin{bmatrix} a \\ b \end{bmatrix} = x\begin{bmatrix} 1 \\ -2 \end{bmatrix} + y\begin{bmatrix} 2 \\ -5 \end{bmatrix} \quad \text{or} \quad \begin{aligned} x + 2y &= a \\ -2x - 5y &= b. \end{aligned}$$

Solve for x and y in terms of a and b to get $x = 5a + 2b$, $y = -2a - b$. Thus

$$(a, b) = (5a + 2b)\mathbf{u}_1 + (-2a - b)\mathbf{u}_2 \quad \text{or,} \quad \text{equivalently,} \quad [(a, b)] = [5a + 2b, -2a - b]^{\mathrm{T}}.$$

Remark: This formula for (a, b) will be used repeatedly below.

255. (A) $G(\mathbf{u}_1) = G(1, -2) = (2 + 6, 4 - 2) = (8, 2)$.
    (B) By Problem 254, $G(\mathbf{u}_1) = (8, 2) = (40 + 4)\mathbf{u}_1 + (-16 - 2)\mathbf{u}_2 = 44\mathbf{u}_1 - 18\mathbf{u}_2$.

256. (A) $G(\mathbf{u}_2) = G(2, -5) = (4 + 15, 8 - 5) = (19, 3)$.
    (B) By Problem 254, $G(\mathbf{u}_2) = (19, 3) = (95 + 6)\mathbf{u}_1 + (-38 - 3)\mathbf{u}_2 = 101\mathbf{u}_1 - 41\mathbf{u}_2$.

257. Write the coordinates of $G(\mathbf{u}_1)$ and $G(\mathbf{u}_2)$ as columns: $[G] = \begin{bmatrix} 44 & 101 \\ -18 & -41 \end{bmatrix}$.

258. We have

$$\begin{bmatrix} a \\ b \end{bmatrix} = x\begin{bmatrix} 1 \\ 3 \end{bmatrix} + y\begin{bmatrix} 2 \\ 5 \end{bmatrix} \quad \text{or} \quad \begin{array}{l} x + 2y = a \\ 3x + 5y = b. \end{array}$$

Solve for x and y in terms of a and b to get $x = 2b - 5a$ and $y = 3a - b$. Thus

$$(a, b) = (-5a + 2b)\mathbf{v}_1 + (3a - b)\mathbf{v}_2.$$

Remark: This formula for (a, b) will be used repeatedly below.

259. (A) $S(\mathbf{v}_1) = S(1, 3) = (6, 3 - 3) = (6, 0)$.
    (B) By Problem 258, $S(\mathbf{v}_1) = (6, 0) = -30\mathbf{v}_1 + 18\mathbf{v}_2$.

260. (A) $S(\mathbf{v}_2) = S(2, 5) = (10, 6 - 5) = (10, 1)$.
    (B) By Problem 258, $S(\mathbf{v}_2) = S(10, 1) = (-50 + 2)\mathbf{v}_1 + (30 - 1)\mathbf{v}_2 = -48\mathbf{v}_1 + 29\mathbf{v}_2$.

261. (A) Write the coordinates of $S(\mathbf{v}_1)$ and $S(\mathbf{v}_2)$ as columns to obtain $[S]_B = \begin{bmatrix} -30 & -48 \\ 18 & 29 \end{bmatrix}$.
    (B) Recall that if $(a, b) \in \mathbf{R}^2$ then $(a, b) = a\mathbf{e}_1 + b\mathbf{e}_2$. Thus

$$\begin{array}{l} S(\mathbf{e}_1) = S(1, 0) = (0, 3) = 0\mathbf{e}_1 + 3\mathbf{e}_2 \\ S(\mathbf{e}_2) = S(0, 1) = (2, -1) = 2\mathbf{e}_1 - \mathbf{e}_2 \end{array} \quad \text{and} \quad [S]_E = \begin{bmatrix} 0 & 2 \\ 3 & -1 \end{bmatrix}.$$

262. (A) $T(\mathbf{v}_1) = T(1, 3) = (3 - 12, 1 + 15) = (-9, 16)$.
    (B) By Problem 258, $T(\mathbf{v}_1) = (-9, 16) = (45 + 32)\mathbf{v}_1 + (-27 - 16)\mathbf{v}_2 = 77\mathbf{v}_1 - 43\mathbf{v}_2$.

263. (A) $T(\mathbf{v}_2) = T(2, 5) = (6 - 20, 2 + 25) = (-14, 27)$.
    (B) By Problem 258, $T(\mathbf{v}_2) = (-14, 27) = (70 + 54)\mathbf{v}_1 + (-42 - 27)\mathbf{v}_2 = 124\mathbf{v}_1 - 69\mathbf{v}_2$.

264. (A) Write the coordinates of $T(\mathbf{v}_1)$ and $T(\mathbf{v}_2)$ as columns to obtain $[T]_B = \begin{bmatrix} 77 & 124 \\ -43 & -69 \end{bmatrix}$.

(B) $\begin{aligned} T(\mathbf{e}_1) &= T(1, 0) = (3, 1) = 3\mathbf{e}_1 + \mathbf{e}_2 \\ T(\mathbf{e}_2) &= T(0, 1) = (-4, 5) = -4\mathbf{e}_1 + 5\mathbf{e}_2 \end{aligned}$ and thus $[T]_E = \begin{bmatrix} 3 & -4 \\ 1 & 5 \end{bmatrix}$.

265. By Question 258, $(a, b) = (-5a + 2b)\mathbf{v}_1 + (3a - b)\mathbf{v}_2$; hence

$$T(\mathbf{v}_1) = \begin{bmatrix} 1 & 2 \\ 3 & 4 \end{bmatrix}\begin{bmatrix} 1 \\ 3 \end{bmatrix} = \begin{bmatrix} 7 \\ 15 \end{bmatrix} = -5\mathbf{v}_1 + 6\mathbf{v}_2$$

$$T(\mathbf{v}_2) = \begin{bmatrix} 1 & 2 \\ 3 & 4 \end{bmatrix}\begin{bmatrix} 2 \\ 5 \end{bmatrix} = \begin{bmatrix} 12 \\ 26 \end{bmatrix} = -8\mathbf{v}_1 + 10\mathbf{v}_2$$

and thus $[T]_B = \begin{bmatrix} -5 & -8 \\ 6 & 10 \end{bmatrix}$.

266. By Theorem 8.2, $[T]_E = A$.

267. By Theorem 8.2, $[T]_E = A$.

268. Write (a, b, c) as a linear combination of $\mathbf{u}_1, \mathbf{u}_2, \mathbf{u}_3$ using unknowns x, y, z:

$$(a, b, c) = x(1, 1, 0) + y(1, 2, 3) + z(1, 3, 5) = (x + y + z, x + 2y + 3z, 3y + 5z)$$

$$\begin{aligned} x + y + z &= a \\ \text{or} \quad x + 2y + 3z &= b \quad \text{or} \\ 3y + 5z &= c \end{aligned} \qquad \begin{aligned} x + y + z &= a \\ y + 2z &= -a + b \\ 3y + 5z &= c \end{aligned} \qquad \text{or} \quad \begin{aligned} x + y + z &= a \\ y + 2z &= -a + b \\ z &= -3a + 3b - c. \end{aligned}$$

Solving for x, y, z in terms of a, b, c yields $x = -a + 2b - c$, $y = 5a - 5b + 2c$, $z = -3a + 3b - c$. Thus

$$(a, b, c,) = (-a + 2b - c)\mathbf{u}_1 + (5a - 5b + 2c)\mathbf{u}_2 + (-3a + 3b - c)\mathbf{u}_3$$

or, equivalently,

$$[(a, b, c)] = [-a + 2b - c, 5a - 5b + 2c, -3a + 3b - c]^\mathrm{T}.$$

Remark: This formula for (a, b, c) will be used repeatedly below.

269. (A) $S(\mathbf{u}_1) = S(1, 1, 0) = (1 + 2 - 0, 2 + 1 + 0, 5 - 1 + 0) = (3, 3, 4)$.
(B) Use Problem 268 to get

$$S(\mathbf{u}_1) = (3, 3, 4) = (-3 + 6 - 4)\mathbf{u}_1 + (15 - 15 + 8)\mathbf{u}_2 + (-9 + 9 - 4)\mathbf{u}_3 = -\mathbf{u}_1 + 8\mathbf{u}_2 - 4\mathbf{u}_3.$$

270. (A) $S(\mathbf{u}_2) = S(1, 2, 3) = (1 + 4 - 9, 2 + 2 + 3, 5 - 2 + 3) = (-4, 7, 6)$.
(B) $S(\mathbf{u}_2) = (-4, 7, 6) = (4 + 14 - 6)\mathbf{u}_1 + (-20 - 35 + 12)\mathbf{u}_2 + (12 + 21 - 6)\mathbf{u}_3 = 12\mathbf{u}_1 - 43\mathbf{u}_2 + 27\mathbf{u}_3$.

271. (A) $S(\mathbf{u}_3) = S(1, 3, 5) = (1 + 6 - 15, 2 + 3 + 5, 5 - 3 + 5) = (-8, 10, 7)$.
  (B) $S(\mathbf{u}_3) = (-8, 10, 7) = (8 + 20 - 7)\mathbf{u}_1 + (-40 - 50 + 14)\mathbf{u}_2 + (24 + 30 - 7)\mathbf{u}_3 = 21\mathbf{u}_1 - 76\mathbf{u}_2 + 47\mathbf{u}_3$.

272. Write the coordinates of $S(\mathbf{u}_1)$, $S(\mathbf{u}_2)$, and $S(\mathbf{u}_3)$ as columns to get

$$[S]_B = \begin{bmatrix} -1 & 12 & 21 \\ 8 & -43 & -76 \\ -4 & 27 & 47 \end{bmatrix}.$$

273. By Remark 8.2, $[S]_E = \begin{bmatrix} 1 & 2 & -3 \\ 2 & 1 & 1 \\ 5 & -1 & 1 \end{bmatrix}$, the coefficients of x, y, z in the definition of S.

274. By Problem 272, $[\mathbf{v}] = [(1, 1, 1)] = [-1 + 2 - 1, 5 - 5 + 2, -3 + 3 - 1]^T = [0, 2, -1]^T$.

275. (A) $S(1, 1, 1) = (1 + 2 - 3, 2 + 1 + 1, 5 - 1 + 1) = (0, 4, 5)$.
  (B) By Problem 272, $[S(\mathbf{v})] = [(0, 4, 5)] = [0 + 8 - 5, 0 - 20 + 10, 0 + 12 - 5]^T = [3, -10, 7]^T$.

276. $[S][\mathbf{v}] = \begin{bmatrix} -1 & 12 & 21 \\ 8 & -43 & -76 \\ -4 & 27 & 47 \end{bmatrix} \begin{bmatrix} 0 \\ 2 \\ -1 \end{bmatrix} = \begin{bmatrix} 0 + 24 - 21 \\ 0 - 86 + 76 \\ 0 + 54 - 47 \end{bmatrix} = \begin{bmatrix} 3 \\ -10 \\ 7 \end{bmatrix} = [S(\mathbf{v})]$.

277. Write (a, b, c) as a linear combination of $\mathbf{w}_1$, $\mathbf{w}_2$, $\mathbf{w}_3$ using unknown scalars x, y, and z:
$(a, b, c) = x(1, 1, 1) + y(1, 1, 0) + z(1, 0, 0) = (x + y + z, x + y, x)$, or $x + y + z = a$, $x + y = b$, $x = c$.

Solve the system of x, y, and z in terms of a, b, and c to find $x = c$, $y = b - c$, $z = a - b$. Thus

$(a, b, c) = c\mathbf{w}_1 + (b - c)\mathbf{w}_2 + (a - b)\mathbf{w}_3$ or, equivalently, $[(a, b, c)] = [c, b - c, a - b]^T$.

Remark: This formula for (a, b, c) will be used repeatedly below.

278. (A) $T(\mathbf{w}_1) = T(1, 1, 1) = (2 + 1, 1 - 4, 3) = (3, -3, 3)$.
  (B) By Problem 277, $T(\mathbf{w}_1) = (3, -3, 3) = 3\mathbf{w}_1 + (-3 - 3)\mathbf{w}_2 + (3 + 3)\mathbf{w}_3 = 3\mathbf{w}_1 - 6\mathbf{w}_2 + 6\mathbf{w}_3$.

279. (A) $T(\mathbf{w}_2) = T(1, 1, 0) = (2 + 0, 1 - 4, 3) = (2, -3, 3)$.
  (B) By Problem 277, $T(\mathbf{w}_2) = T(2, -3, 3) = 3\mathbf{w}_1 + (-3 - 3)\mathbf{w}_2 + (2 + 3)\mathbf{w}_3 = 3\mathbf{w}_1 - 6\mathbf{w}_2 + 5\mathbf{w}_3$.

280. (A) $T(\mathbf{w}_3) = T(1, 0, 0) = (0 + 0, 1 - 0, 3) = (0, 1, 3)$.
  (B) By Problem 277, $T(\mathbf{w}_3) = (0, 1, 3) = 3\mathbf{w}_1 + (1 - 3)\mathbf{w}_2 + (0 - 1)\mathbf{w}_3 = 3\mathbf{w}_1 - 2\mathbf{w}_2 - \mathbf{w}_3$.

281. (A) Write the coordinates of $T(\mathbf{w}_1)$, $T(\mathbf{w}_2)$, $T(\mathbf{w}_3)$ as columns to obtain

$$[T]_B = \begin{bmatrix} 3 & 3 & 3 \\ -6 & -6 & -2 \\ 6 & 5 & -1 \end{bmatrix}.$$

(B) By Remark 8.2, $[T]_E = \begin{bmatrix} 0 & 2 & 1 \\ 1 & -4 & 0 \\ 3 & 0 & 0 \end{bmatrix}$.

282. $T(\mathbf{v}) = T(a, b, c) = (2b + c, a - 4b, 3a) = 3a\mathbf{w}_1 + (-2a - 4b)\mathbf{w}_2 + (-a + 6b + c)\mathbf{w}_3$
or, equivalently, $[T(\mathbf{v})] = [3a, -2a - 4b, - a + 6b + c]^{\mathrm{T}}$.

283. $[T][\mathbf{v}] = \begin{bmatrix} 3 & 3 & 3 \\ -6 & -6 & -2 \\ 6 & 5 & -1 \end{bmatrix} \begin{bmatrix} c \\ b - c \\ a - b \end{bmatrix} = \begin{bmatrix} 3a \\ -2a - 4b \\ -a + 6b + c \end{bmatrix} = [T(\mathbf{v})]$.

284. Write (a, b) as a linear combination of $\mathbf{u}_1$ and $\mathbf{u}_2$ using unknowns x and y: $(a, b) = x(1, 1) + y(1, 2)$, or $x + y = a$, $x + 2y = b$. Solve for x and y to get $x = 2a - b$, $y = -a + b$. Thus

$$(a, b) = (2a - b)\mathbf{u}_1 + (-a + b)\mathbf{u}_2 \qquad \text{or, equivalently,} \qquad [(a, b)] = [2a - b, -a + b]^{\mathrm{T}}.$$

Remark: This formula for (a, b) will be used repeatedly.

285. (A) $S(\mathbf{u}_1) = S(1, 1) = (3, 4) = 2\mathbf{u}_1 + \mathbf{u}_2$ and thus $[S] = \begin{bmatrix} 2 & 6 \\ 1 & -1 \end{bmatrix}$.
$\qquad\quad S(\mathbf{u}_2) = S(1, 2) = (5, 4) = 6\mathbf{u}_1 - \mathbf{u}_2$

(B) $T(\mathbf{u}_1) = T(1, 1) = (1, 4) = -2\mathbf{u}_1 + 3\mathbf{u}_2$ and thus $[T] = \begin{bmatrix} -2 & -3 \\ 3 & 5 \end{bmatrix}$.
$\qquad\quad T(\mathbf{u}_2) = T(1, 2) = (2,7) = -3\mathbf{u}_1 + 5\mathbf{u}_2$

286. (A) $(S + T)(\mathbf{u}_1) = S(\mathbf{u}_1) + T(\mathbf{u}_1) = (2\mathbf{u}_1 + \mathbf{u}_2) + (-2\mathbf{u}_1 + 3\mathbf{u}_2) = 0\mathbf{u}_1 + 4\mathbf{u}_2$.
(B) $(S + T)(\mathbf{u}_2) = S(\mathbf{u}_2) + T(\mathbf{u}_2) = (6\mathbf{u}_1 - \mathbf{u}_2) + (-3\mathbf{u}_1 + 5\mathbf{u}_2) = 3\mathbf{u}_1 + 4\mathbf{u}_2$.

287. (A) Write the coordinates of $(S + T)(\mathbf{u}_1)$ and $(S + T)(\mathbf{u}_2)$ as columns: $[S + T] = \begin{bmatrix} 0 & 3 \\ 4 & 4 \end{bmatrix}$.

(B) $[S] + [T] = \begin{bmatrix} 2 & 6 \\ 1 & -1 \end{bmatrix} + \begin{bmatrix} -2 & -3 \\ 3 & 5 \end{bmatrix} = \begin{bmatrix} 0 & 3 \\ 4 & 4 \end{bmatrix} = [S + T]$.

288. (A) $(3T)(\mathbf{u}_1) = 3T(\mathbf{u}_1) = 3(-2\mathbf{u}_1 + 3\mathbf{u}_2) = -6\mathbf{u}_1 + 9\mathbf{u}_2$.
(B) $(3T)(\mathbf{u}_2) = 3T(\mathbf{u}_2) = 3(-3\mathbf{u}_1 + 5\mathbf{u}_2) = -9\mathbf{u}_1 + 15\mathbf{u}_2$.

289. (A) Write the coordinates of $(3T)(\mathbf{u}_1)$ and $(3T)(\mathbf{u}_2)$ as columns: $[3T] = \begin{bmatrix} -6 & -9 \\ 9 & 15 \end{bmatrix}$.

(B) $3[T] = 3\begin{bmatrix} -2 & -3 \\ 3 & 5 \end{bmatrix} = \begin{bmatrix} -6 & -9 \\ 9 & 15 \end{bmatrix} = [3T]$.

290. (A) $(S \circ T)(\mathbf{u}_1) = S(T(\mathbf{u}_1)) = S(T(1, 1)) = S(1, 4) = (9, 4) = (18 - 4)\mathbf{u}_1 + (-9 + 4)\mathbf{u}_2 = 14\mathbf{u}_1 - 5\mathbf{u}_2$.

(B) $(S \circ T)(\mathbf{u}_2) = S(T(\mathbf{u}_2)) = S(T(1, 2)) = S(2, 7) = (16, 8) = (32 - 8)\mathbf{u}_1 + (-16 + 8)\mathbf{u}_2 = 24\mathbf{u}_1 - 8\mathbf{u}_2$.

291. (A) Write the coordinates of $(S \circ T)(\mathbf{u}_1)$ and $(S \circ T)(\mathbf{u}_2)$ as columns: $[S \circ T] = \begin{bmatrix} 14 & 24 \\ -5 & -8 \end{bmatrix}$.

(B) $[S][T] = \begin{bmatrix} 2 & 6 \\ 1 & -1 \end{bmatrix}\begin{bmatrix} -2 & -3 \\ 3 & 5 \end{bmatrix} = \begin{bmatrix} -4+18 & -6+30 \\ -2-3 & -3-5 \end{bmatrix} = \begin{bmatrix} 14 & 24 \\ -5 & -8 \end{bmatrix} = [S \circ T]$.

292. The mapping is one-to-one since a linear mapping is completely determined by its values on a basis. The mapping is onto since each matrix $M \in \mathbf{M}$ is the image of the linear operator

$$F(\mathbf{e}_i) = \sum_{j=1}^{n} m_{ij}\mathbf{e}_j \qquad i = 1, \dots, n$$

where $[m_{ij}]$ is the transpose of the matrix M.

293. $\begin{bmatrix} a \\ b \end{bmatrix} = x\begin{bmatrix} 1 \\ 2 \end{bmatrix} + y\begin{bmatrix} 2 \\ 3 \end{bmatrix}$ or $\begin{array}{l} x + 2y = a \\ 2x + 3y = b. \end{array}$

The solution is $x = -3a + 2b$, $y = 2a - b$. Thus

$(a, b) = (-3a + 2b)\mathbf{v}_1 + (2a - b)\mathbf{v}_2$, or $[(a, b)]_{B_2} = [-3a + 2b, 2a - b]^{\mathrm{T}}$.

294. $F(\mathbf{u}_1) = F(1, 1, 0) = (2 + 3 + 0, 4 - 1 + 0) = (5, 3) = (-15 + 6)\mathbf{v}_1 + (10 - 3)\mathbf{v}_2 = -9\mathbf{v}_1 + 7\mathbf{v}_2$.

295. $F(\mathbf{u}_2) = F(1, 2, 3) = (2 + 6 - 3, 4 - 2 + 6) = (5, 8) = (-15 + 16)\mathbf{v}_1 + (10 - 8)\mathbf{v}_2 = \mathbf{v}_1 + 2\mathbf{v}_2$.

296. $F(\mathbf{u}_3) = F(1, 3, 5) = (2 + 9 - 5, 4 - 3 + 10) = (6, 11) = (-18 + 22)\mathbf{v}_1 + (12 - 11)\mathbf{v}_2 = 4\mathbf{v}_1 + \mathbf{v}_2$.

297. Write the coordinates of $F(\mathbf{u}_1)$, $F(\mathbf{u}_2)$, $F(\mathbf{u}_3)$ as columns: $[F] = \begin{bmatrix} -9 & 1 & 4 \\ 7 & 2 & 1 \end{bmatrix}$.

298. By Problem 268, $[(a, b, c)]_{B_1} = [-a + 2b - c, 5a - 5b + 2c, -3a + 3b - c]^T$, Hence

$$[v]_{B_1} = [-2 + 10 + 3, 10 - 25 - 6, -6 + 15 + 3]^T = [11, -21, 12]^T.$$

299. **(A)** $F(v) = F(2, 5, -3) = (4 + 15 + 3, 8 - 5 - 6) = (22, -3).$
(B) By Problem 293, $[F(v)]_{B_2} = [(22, -3)]_{B_2} = [-66 - 6, 44 + 3]^T = [-72, 47]^T.$

300. $[F][v]_{B_1} = \begin{bmatrix} -9 & 1 & 4 \\ 7 & 2 & 1 \end{bmatrix} \begin{bmatrix} 11 \\ -21 \\ 12 \end{bmatrix} = \begin{bmatrix} -99 - 21 + 48 \\ 77 - 42 + 12 \end{bmatrix} = \begin{bmatrix} -72 \\ 47 \end{bmatrix} = [F(v)]_{B_2}$

301. Simply write the coefficients of x and y as rows to get $[T]_E = \begin{bmatrix} 3 & -5 \\ 2 & 7 \end{bmatrix}$.

302. Using $(a, b) = (-5a + 2b)v_1 + (2a - b)v_2$, we get

$$T(v_1) = T(1, 3) = (3 - 15, 2 + 23) = (-12, 25) = (60 + 46)v_1 + (-36 - 23)v_2 = 106v_1 - 59v_2$$
$$T(v_2) = T(2, 5) = (6 - 25, 4 + 35) = (-19, 39) = (95 + 78)v_1 + (-57 - 39)v_2 = 173v_1 - 96v_2.$$

Write the coordinates of $T(v_1)$ and $T(v_2)$ as columns to get $[T]_S = \begin{bmatrix} 106 & 173 \\ -59 & -96 \end{bmatrix}$.

303. Write v_1 and v_2 as columns to get $P = \begin{bmatrix} 1 & 2 \\ 3 & 5 \end{bmatrix}$. Since $|P| = -1$, $P^{-1} = \begin{bmatrix} -5 & 2 \\ 3 & -1 \end{bmatrix}$.

304. $[T]_S = P^{-1}[T]_E P = \begin{bmatrix} -5 & 2 \\ 3 & -1 \end{bmatrix} \begin{bmatrix} 3 & -5 \\ 2 & 7 \end{bmatrix} \begin{bmatrix} 1 & 2 \\ 3 & 5 \end{bmatrix} = \begin{bmatrix} 106 & 173 \\ -59 & -96 \end{bmatrix}$.

305. We have $[L]_E = \begin{bmatrix} 0 & 2 \\ 3 & -1 \end{bmatrix}$ where E is the usual basis of $\mathbf{R}^2$. Thus

$$[L]_S = P^{-1}[L]_E P = \begin{bmatrix} -5 & 2 \\ 3 & -1 \end{bmatrix} \begin{bmatrix} 0 & 2 \\ 3 & -1 \end{bmatrix} \begin{bmatrix} 1 & 2 \\ 3 & 5 \end{bmatrix} = \begin{bmatrix} -30 & -48 \\ 18 & 29 \end{bmatrix}.$$

306. Write the basis vectors as columns to get $P = \begin{bmatrix} 1 & 3 \\ 4 & 10 \end{bmatrix}$. Use the formula for the inverse of a 2-square matrix to get $P^{-1} = \begin{bmatrix} -5 & \frac{3}{2} \\ 2 & -\frac{1}{2} \end{bmatrix}$. Thus

$$B = P^{-1}AP = \begin{bmatrix} -5 & \frac{3}{2} \\ 2 & -\frac{1}{2} \end{bmatrix} \begin{bmatrix} 5 & -7 \\ 2 & 3 \end{bmatrix} \begin{bmatrix} 1 & 3 \\ 4 & 10 \end{bmatrix} = \begin{bmatrix} 136 & 329 \\ -53 & -128 \end{bmatrix}.$$

307. By Theorem 8.7,

$$B = P^{-1}AP = \begin{bmatrix} -2 & -7 & 5 \\ 2 & 4 & -3 \\ -1 & -1 & 1 \end{bmatrix} \begin{bmatrix} 2 & 3 & -4 \\ 4 & -6 & 3 \\ 1 & 4 & -2 \end{bmatrix} \begin{bmatrix} 1 & 2 & 1 \\ 1 & 3 & 4 \\ 2 & 5 & 6 \end{bmatrix} = \begin{bmatrix} -17 & -1 & 59 \\ 7 & -6 & -43 \\ 0 & 6 & 17 \end{bmatrix}.$$

308. The matrix P is the change-of-basis matrix from the usual basis of $\mathbf{R}^3$ to S:

$$B = P^{-1}AP = \begin{bmatrix} 1 & -1 & 0 \\ 0 & 1 & -1 \\ 0 & 0 & 1 \end{bmatrix}\begin{bmatrix} 1 & 3 & 5 \\ 2 & 4 & 6 \\ 7 & 8 & 9 \end{bmatrix}\begin{bmatrix} 1 & 1 & 1 \\ 0 & 1 & 1 \\ 0 & 0 & 1 \end{bmatrix} = \begin{bmatrix} -1 & -2 & -3 \\ -5 & -9 & -12 \\ 7 & 15 & 24 \end{bmatrix}.$$

309. The identity matrix I is invertible and $I = I^{-1}$. Since $A = I^{-1}AI$, A is similar to A.

310. Since A is similar to B, there exists an invertible matrix P such that $A = P^{-1}BP$. Hence $B = PAP^{-1} = (P^{-1})^{-1}AP^{-1}$ and P^{-1} is invertible. Thus B is similar to A.

311. Since A is similar to B, there exists an invertible matrix P such that $A = P^{-1}BP$, and since B is similar to C, there exists an invertible matrix Q such that $B = Q^{-1}CQ$. Hence $A = P^{-1}BP = P^{-1}(Q^{-1}CQ)P = (QP)^{-1}C(QP)$ and QP is invertible. Thus A is similar to C.

312. By Remark 8.2, $A = \begin{bmatrix} 3 & -7 \\ 4 & 8 \end{bmatrix}$.

313. **(A)** $\text{tr}(F) = \text{tr}(A) = 11$.
(B) $\det(F) = \det(A) = 24 + 28 = 52$.

314. By Remark 8.2, $B = \begin{bmatrix} 2 & 0 & -1 \\ 1 & 2 & -4 \\ 3 & -3 & 1 \end{bmatrix}$.

315. **(A)** $\text{tr}(T) = \text{tr}(B) = 2 + 2 + 1 = 5$.
(B) $\det(T) = \det(B) = 4 + 0 + 3 + 6 - 24 - 0 = -11$.

Chapter 9: Inner Product Spaces

316. $\langle \mathbf{0}, \mathbf{v} \rangle = \langle 0\mathbf{v}, \mathbf{v} \rangle = 0\langle \mathbf{v}, \mathbf{v} \rangle = 0$. Also, $\langle \mathbf{v}, \mathbf{0} \rangle = \langle \mathbf{0}, \mathbf{v} \rangle = 0$.

317. By [RIP$_1$] and [RIP$_2$], we have

$$\langle \mathbf{u}, \mathbf{v}_1 + \mathbf{v}_2 \rangle = \langle \mathbf{v}_1 + \mathbf{v}_2, \mathbf{u} \rangle = \langle \mathbf{v}_1, \mathbf{u} \rangle + \langle \mathbf{v}_2, \mathbf{u} \rangle = \langle \mathbf{u}, \mathbf{v}_1 \rangle + \langle \mathbf{u}, \mathbf{v}_2 \rangle.$$

318. $\langle \mathbf{u}, k\mathbf{v} \rangle = \langle k\mathbf{v}, \mathbf{u} \rangle = k\langle \mathbf{v}, \mathbf{u} \rangle = k\langle \mathbf{u}, \mathbf{v} \rangle$.

319. **(A)** Multiply corresponding components and add to get $\mathbf{u} \cdot \mathbf{v} = 2 - 6 + 20 = 16$.
(B) $\mathbf{u} \cdot \mathbf{w} = 4 + 4 - 12 = -4$.
(C) $\mathbf{v} \cdot \mathbf{w} = 8 - 6 - 15 = -13$.

320. First find $\mathbf{u} + \mathbf{v} = (3, -1, 9)$. Then $(\mathbf{u} \cdot \mathbf{v}) \cdot \mathbf{w} = 12 - 2 - 27 = -17$. Alternatively, using [RIP$_1$], $(\mathbf{u} + \mathbf{v}) \cdot \mathbf{w} = \mathbf{u} \cdot \mathbf{w} + \mathbf{v} \cdot \mathbf{w} = -4 - 13 = -17$.

321. (A) First find $\|\mathbf{u}\|^2$ by squaring the components of $\mathbf{u}$ and adding:
$\|\mathbf{u}\|^2 = 1^2 + 2^2 + 4^2 = 1 + 4 + 16 = 21$. Then $\|\mathbf{u}\| = \sqrt{21}$.

   (B) $\|\mathbf{v}\|^2 = 4 + 9 + 25 = 38$ and so $\|\mathbf{v}\| = \sqrt{38}$.

   (C) First find $\mathbf{u} + \mathbf{v} = (3, -1, 9)$. Hence $\|\mathbf{u} + \mathbf{v}\|^2 = 9 + 1 + 81 = 91$. Thus $\|\mathbf{u} + \mathbf{v}\| = \sqrt{91}$.

322. (A) $\langle \mathbf{f}, \mathbf{g} \rangle = \displaystyle\int_0^1 (t+2)(3t-2)\, dt = \int_0^1 (3t^2 + 4t - 4)\, dt = [t^3 + 2t^2 - 4t]_0^1 = -1$.

   (B) $\langle \mathbf{f}, \mathbf{h} \rangle = \displaystyle\int_0^1 (t+2)(t^2 - 2t - 3)\, dt = \left[\dfrac{t^4}{4} - \dfrac{7t^2}{2} - 6t \right]_0^1 = -\dfrac{37}{4}$.

323. (A) $\langle \mathbf{f}, \mathbf{f} \rangle = \displaystyle\int_0^1 (t+2)(t+2)\, dt = \dfrac{19}{3}$ and $\|\mathbf{f}\| = \sqrt{\langle \mathbf{f}, \mathbf{f} \rangle} = \sqrt{\dfrac{19}{3}} = \dfrac{1}{3}\sqrt{57}$.

   (B) $\langle \mathbf{g}, \mathbf{g} \rangle = \displaystyle\int_0^1 (3t-2)(3t-2) = 1$; hence $\|\mathbf{g}\| = \sqrt{1} = 1$.

324. $\|\mathbf{u} + \mathbf{v}\|^2 = \langle \mathbf{u} + \mathbf{v}, \mathbf{u} + \mathbf{v} \rangle = \langle \mathbf{u}, \mathbf{u} \rangle + \langle \mathbf{u}, \mathbf{v} \rangle + \langle \mathbf{v}, \mathbf{u} \rangle + \langle \mathbf{v}, \mathbf{v} \rangle =$
$\langle \mathbf{u}, \mathbf{u} \rangle + \langle \mathbf{u}, \mathbf{v} \rangle + \langle \mathbf{u}, \mathbf{v} \rangle + \langle \mathbf{v}, \mathbf{v} \rangle = \|\mathbf{u}\|^2 + 2\langle \mathbf{u}, \mathbf{v} \rangle + \|\mathbf{v}\|^2$

325. $\|\mathbf{u} - \mathbf{v}\|^2 = \langle \mathbf{u} - \mathbf{v}, \mathbf{u} - \mathbf{v} \rangle = \langle \mathbf{u}, \mathbf{u} \rangle - \langle \mathbf{u}, \mathbf{v} \rangle - \langle \mathbf{v}, \mathbf{u} \rangle + \langle \mathbf{v}, \mathbf{v} \rangle =$
$\langle \mathbf{u}, \mathbf{u} \rangle - \langle \mathbf{u}, \mathbf{v} \rangle - \langle \mathbf{u}, \mathbf{v} \rangle + \langle \mathbf{v}, \mathbf{v} \rangle = \|\mathbf{u}\|^2 - 2\langle \mathbf{u}, \mathbf{v} \rangle + \|\mathbf{v}\|^2$

326. Subtract the equation in Question 325 from the equation in Question 324 to get $\|\mathbf{u} + \mathbf{v}\|^2 - \|\mathbf{u} - \mathbf{v}\|^2 = 4 \langle \mathbf{u}, \mathbf{v} \rangle$. Dividing by 4 gives us our result.

327. Suppose $\hat{\mathbf{v}} = k\mathbf{v}$ where $k > 0$ and $\|\hat{\mathbf{v}}\| = 1$. Then $1 = \|\hat{\mathbf{v}}\|^2 = \langle k\mathbf{v}, k\mathbf{v} \rangle = k^2 \langle \mathbf{v}, \mathbf{v} \rangle = k^2 \|\mathbf{v}\|^2$.
Since k is positive, we get $k = 1/\|\mathbf{v}\|$.

328. (A) Note that $\langle \mathbf{u}, \mathbf{u} \rangle$ is the sum of the squares of the entries of $\mathbf{u}$; that is, $\langle \mathbf{u}, \mathbf{u} \rangle = 2^2 + 1^2 + (-1)^2 = 6$. Hence divide $\mathbf{u}$ by $\|\mathbf{u}\| = \sqrt{\langle \mathbf{u}, \mathbf{u} \rangle} = \sqrt{6}$ to obtain the required unit vector:
$\hat{\mathbf{u}} = \hat{\mathbf{u}}/\|\hat{\mathbf{u}}\| = (2/\sqrt{6}, 1/\sqrt{6}, -1/\sqrt{6})$.

   (B) First multiply $\mathbf{v}$ by 12 to "clear" fractions, obtaining $12\mathbf{v} = (6, 8, -3)$. We have $\langle 12\mathbf{v}, 12\mathbf{v} \rangle = 6^2 + 8^2 + (-3)^2 = 109$. Then the required unit vector is $\hat{\mathbf{v}} = 12\hat{\mathbf{v}}/\|12\hat{\mathbf{v}}\| = (6/\sqrt{109}, 8/\sqrt{109}, -3/\sqrt{109})$.

329. $\mathbf{u} - \mathbf{v} = (3, 10)$ and $\|\mathbf{u} - \mathbf{v}\|^2 = 9 + 100 = 109$. Hence $d(\mathbf{u}, \mathbf{v}) = \sqrt{109}$.

330. First find $\mathbf{u} - \mathbf{v} = (5 - 1, 5 - 2, 8 - 3, 8 - 4) = (4, 3, 5, 4)$. Then find $\|\mathbf{u} - \mathbf{v}\|^2 = 4^2 + 3^2 + 5^2 + 4^2 = 16 + 9 + 25 + 16 = 66$. Hence $d(\mathbf{u}, \mathbf{v}) = \sqrt{66}$.

331. We have $f(t) - g(t) = -2t + 4$. Then $\|\mathbf{f} - \mathbf{g}\|^2 = \langle \mathbf{f} - \mathbf{g}, \mathbf{f} - \mathbf{g} \rangle = \displaystyle\int_0^1 (-2t + 4)(-2t + 4)\, dt = \int_0^1 (4t^2 - 16t + 16) = [\tfrac{4}{3}t^3 - 8t^2 + 16t]_0^1 = \tfrac{28}{3}$.
   Hence $d(\mathbf{f}, \mathbf{g}) = \sqrt{\tfrac{28}{3}} = \tfrac{2}{3}\sqrt{21}$.

332. (A) Compute $\langle \mathbf{u}, \mathbf{v} \rangle = 2 - 3 + 10 = 9, \|\mathbf{u}\|^2 = 1 + 9 + 4 = 14, \|\mathbf{u}\|^2 = 4 + 1 + 25 = 30.$
Thus

$$\cos\theta = \frac{9}{\sqrt{14}\sqrt{30}} = \frac{9}{\sqrt{105}}.$$

(B) Compute $\langle \mathbf{u}, \mathbf{v} \rangle = -10 + 3 = -7, \|\mathbf{u}\|^2 = 25 + 1 = 26, \|\mathbf{v}\|^2 = 4 + 9 = 13.$ Thus

$$\cos\theta = \frac{-7}{\sqrt{13}\sqrt{26}} = -\frac{7}{13\sqrt{2}}.$$

Since $\cos\theta$ is negative, θ lies in the second quadrant.

333. Compute

$$\langle \mathbf{f}, \mathbf{g} \rangle = \int_0^1 (2t^3 - t^2)\, dt = \left[\frac{t^4}{2} - \frac{t^3}{3} \right]_0^1 = \frac{1}{2} - \frac{1}{3} = \frac{1}{6}$$

$$\|\mathbf{f}\|^2 = \langle \mathbf{f}, \mathbf{f} \rangle = \int_0^1 (4t^2 - 4t + 1)\, dt = \frac{1}{3}$$

$$\|\mathbf{g}\|^2 = \langle \mathbf{g}, \mathbf{g} \rangle = \int_0^1 t^4 \, dt = \frac{1}{5}.$$

Thus

$$\cos\theta = \frac{\frac{1}{6}}{(1/\sqrt{3})(1/\sqrt{5})} = \frac{\sqrt{15}}{6}.$$

334. (A) If $\mathbf{v} \neq \mathbf{0}$, then $\langle \mathbf{v}, \mathbf{v} \rangle > 0$ and hence $\|\mathbf{v}\| = \sqrt{\langle \mathbf{v}, \mathbf{v} \rangle} > 0$. If $\mathbf{v} = \mathbf{0}$, then $\langle \mathbf{0}, \mathbf{0} \rangle = 0$ and so $\|\mathbf{0}\| = \sqrt{0} = 0$.
(B) We have $\|k\mathbf{v}\|^2 = \langle k\mathbf{v}, k\mathbf{v} \rangle = k^2 \langle \mathbf{v}, \mathbf{v} \rangle = k^2 \|\mathbf{v}\|^2$. Taking the square root of both sides gives [N₂].
(C) Using the Cauchy–Schwarz inequality, we obtain $\|\mathbf{u} + \mathbf{v}\|^2 = \langle \mathbf{u} + \mathbf{v}, \mathbf{u} + \mathbf{v} \rangle = \langle \mathbf{u}, \mathbf{u} \rangle + \langle \mathbf{u}, \mathbf{v} \rangle + \langle \mathbf{u}, \mathbf{v} \rangle + \langle \mathbf{v}, \mathbf{v} \rangle \leq \|\mathbf{u}\|^2 + 2\|\mathbf{u}\|\|\mathbf{v}\| + \|\mathbf{v}\|^2 = (\|\mathbf{u}\| + \|\mathbf{v}\|)^2$. Taking the square root of both sides yields [N₃].

335. Clearly, $\mathbf{0} \in W^\perp$. Now suppose $\mathbf{u}, \mathbf{v} \in W^\perp$. Then for any $a, b \in K$ and any $\mathbf{w} \in W$, $\langle a\mathbf{u} + b\mathbf{v}, \mathbf{w} \rangle = a\langle \mathbf{u}, \mathbf{w} \rangle + b\langle \mathbf{u}, \mathbf{w} \rangle = a \cdot 0 + b \cdot 0 = 0$. Thus $a\mathbf{u} + b\mathbf{v} \in W^\perp$ and therefore W is a subspace of V.

336. Note that $\mathbf{u}^\perp$ consists of all vectors (x, y, z) such that $\langle (x, y, z), (1, 3, -4) \rangle = 0$, or $x + 3y - 4z = 0$. The free variables are y and z. Set $y = -1, z = 0$ to obtain the solution $\mathbf{w}_1 = (3, -1, 0)$, and set $y = 0, z = 1$ to obtain the solution $\mathbf{w}_2 = (4, 0, 1)$. The vectors $\mathbf{w}_1$ and $\mathbf{w}_2$ form a basis for the solution space of the equation and hence a basis for $\mathbf{u}^\perp$.

337. We seek all vectors $\mathbf{w} = (x, y, z, s, t)$ such that

$$\langle \mathbf{w}, \mathbf{u} \rangle = x + 2y + 3z - s + 2t = 0$$
$$\langle \mathbf{w}, \mathbf{v} \rangle = 2x + 4y + 7z + 2s - t = 0.$$

Eliminating x from the second equation, we find the equivalent system

$$x + 2y + 3z - s + 2t = 0$$
$$z + 4s - 5t = 0.$$

The free variables are y, s, and t. Set $y = -1$, $s = 0$, $t = 0$ to obtain the solution $\mathbf{w}_1 = (2, -1, 0, 0, 0)$. Set $y = 0$, $s = 1$, $t = 0$ to find the solution $\mathbf{w}_2 = (13, 0, -4, 1, 0)$. Set $y = 0$, $s = 0$, $t = 1$ to obtain the solution $\mathbf{w}_3 = (-17, 0, 5, 0, 1)$. The set $\{\mathbf{w}_1, \mathbf{w}_2, \mathbf{w}_3\}$ is a basis of $W^{\perp}$.

338. Each solution vector $\mathbf{v} = (x_1, x_2, \ldots, x_n)$ is orthogonal to each row of A. Thus W is the orthogonal complement of the row space of A.

339. $\langle \mathbf{u}, \mathbf{v} \rangle = 3 + 8 - 3 - 8 = 0$, $\langle \mathbf{u}, \mathbf{w} \rangle = 3 - 4 - 3 + 4 = 0$, $\langle \mathbf{v}, \mathbf{w} \rangle = 9 - 8 + 1 - 2 = 0$. Each pair of vectors is orthogonal; hence S is orthogonal.

340. Divide each vector in S by its length. First find $\|\mathbf{u}\|^2 = 1 + 4 + 9 + 16 = 30$, $\|\mathbf{v}\|^2 = 9 + 16 + 1 + 4 = 30$, $\|\mathbf{w}\|^2 = 9 + 4 + 1 + 1 = 15$. Then $\hat{\mathbf{u}} = (1/\sqrt{30}, 2/\sqrt{30}, -3/\sqrt{30}, 4/\sqrt{30})$, $\hat{\mathbf{v}} = (3/\sqrt{30}, 4/\sqrt{30}, 1/\sqrt{30}, -2/\sqrt{30})$, $\hat{\mathbf{w}} = (3/\sqrt{15}, -2/\sqrt{30}, 1/\sqrt{15}, 1/\sqrt{15})$ form the desired orthonormal set of vectors.

341. We have $\langle e_1, e_2 \rangle = 0$, $\langle e_1, e_3 \rangle = 0$, and $\langle e_2, e_3 \rangle = 0$. Thus E is orthogonal. Furthermore, $\langle e_1, e_1 \rangle = 1$, $\langle e_2, e_2 \rangle = 1$, and $\langle e_3, e_3 \rangle = 1$. Thus E is an orthonormal basis of $\mathbf{R}^3$.

Remark: The above is true in general, i.e., the usual basis of $\mathbf{R}^n$ is orthonormal for every n.

342. **(A)** $\mathbf{u}_1 \cdot \mathbf{u}_2 = 2 + 2 - 4 = 0$, $\mathbf{u}_1 \cdot \mathbf{u}_3 = 3 - 4 + 1 = 0$, $\mathbf{u}_2 \cdot \mathbf{u}_3 = 6 - 2 - 4 = 0$. Thus S is orthogonal.
    (B) Since S is orthogonal it is linearly independent, and any three linearly independent vectors form a basis for $\mathbf{R}^3$.

343. First form the equation

$$(3, 4, 5) = x(1, 2, 1) + y(2, 1, -4) + z(3, -2, 1). \tag{1}$$

Take the inner product of (1) with respect to $\mathbf{u}_1$ to get $(3, 4, 5) \cdot (1, 2, 1) = x(1, 2, 1) \cdot (1, 2, 1)$, or $16 = 6x$, or $x = \frac{8}{3}$. Take the inner product of (1) with respect to $\mathbf{u}_2$ to get $(3, 4, 5) \cdot (2, 1, -4) = y(2, 1, -4) \cdot (2, 1, -4)$, or $-10 = 21y$, or $y = -\frac{10}{21}$. Take the inner product of (1) with respect to $\mathbf{u}_3$ to get $(3, 4, 5) \cdot (3, -2, 1) = z(3, -2, 1) \cdot (3, -2, 1)$, or $6 = 14z$, or $z = \frac{3}{7}$. Thus $\mathbf{w} = \frac{8}{3}\mathbf{u}_1 - \frac{10}{21}\mathbf{u}_2 + \frac{3}{7}\mathbf{u}_3$.

344. $\|\mathbf{u}_1\|^2 = 1 + 4 + 1 = 6$, $\|\mathbf{u}_2\|^2 = 4 + 1 + 16 = 21$, $\|\mathbf{u}_3\|^2 = 9 + 4 + 1 = 14$.
    Thus $\hat{\mathbf{u}}_1 = (1/\sqrt{6}, 2/\sqrt{6}, 1/\sqrt{6})$, $\hat{\mathbf{u}}_2 = (2/\sqrt{21}, 1/\sqrt{21}, -4/\sqrt{21})$, $\hat{\mathbf{u}}_3 = (3/\sqrt{14}, -2/\sqrt{14}, 1/\sqrt{14})$ form the desired orthonormal basis of $\mathbf{R}^3$.

345. Find a nonzero solution of $x + 2y + 3z = 0$; say, $\mathbf{v}_1 = (1, 1, -1)$. Now find a nonzero solution to the system $x + 2y + 3z = 0$, $x + y - z = 0$ to obtain $\mathbf{v}_2 = (5, -4, 1)$. (Alternatively, $\mathbf{v}_2$ can be obtained by taking the cross product $\mathbf{w} \times \mathbf{v}_1$.) Then $\{\mathbf{v}_1, \mathbf{v}_2\}$ is an orthogonal basis for $\mathbf{w}^{\perp}$.

346. Normalize the orthogonal basis obtained above: $\|\mathbf{v}_1\|^2 = 1 + 1 + 1 = 3$, $\|\mathbf{v}_2\|^2 = 25 + 16 + 1 = 42$.

Thus $\hat{\mathbf{v}}_1 = (1/\sqrt{3}, 1/\sqrt{3}, -1/\sqrt{3})$ and $\mathbf{v}_2 = (5/\sqrt{42}, -4/\sqrt{42}, 1/\sqrt{42})$ form an orthonormal basis of $\mathbf{w}^{\perp}$.

347. Since $\mathbf{u}_i \neq \mathbf{0}$ and $a_i \neq 0$, we have $a_i\mathbf{u}_i \neq \mathbf{0}$. Also, for $i \neq j$, $\langle \mathbf{u}_i, \mathbf{u}_j \rangle = 0$ and hence $\langle a_i\mathbf{u}_i, a_j\mathbf{u}_j \rangle = a_i a_j \langle \mathbf{u}_i, \mathbf{u}_j \rangle = a_i a_j \cdot 0 = 0$. Thus $\{a_i\mathbf{u}_i\}$ is orthogonal.

348. By Theorem 9.5, $P = \begin{bmatrix} 1/\sqrt{10} & 3/\sqrt{10} \\ -3/\sqrt{10} & 1/\sqrt{10} \end{bmatrix}$ or $\begin{bmatrix} 1/\sqrt{10} & 3/\sqrt{10} \\ 3/\sqrt{10} & -1/\sqrt{10} \end{bmatrix}$.

349. First find a nonzero vector $\mathbf{w}_2 = (x, y, z)$ that is orthogonal to $\mathbf{u}_1$ or, equivalently, to $\mathbf{w}_1 = 3\mathbf{u}_1 = (1, 2, 2)$. We have

$$\langle \mathbf{w}_1, \mathbf{w}_2 \rangle = (1, 2, 2) \cdot (x, y, z) = 0, \quad \text{or} \quad x + 2y + 2z = 0.$$

One solution is $\mathbf{w}_2 = (0, 1, -1)$. Next find a nonzero vector $\mathbf{w}_3 = (x, y, z)$ that is orthogonal to both $\mathbf{w}_1$ and $\mathbf{w}_2$. We have

$$\langle \mathbf{w}_1, \mathbf{w}_3 \rangle = (1, 2, 2) \cdot (x, y, z) = x + 2y + 2z = 0$$
$$\langle \mathbf{w}_2, \mathbf{w}_3 \rangle = (0, 1, -1) \cdot (x, y, z) = y - z = 0.$$

Set $z = -1$ and find the solution $\mathbf{w}_3 = (4, -1, -1)$. Normalize $\mathbf{w}_2$ and $\mathbf{w}_3$ to obtain, respectively,

$$\mathbf{u}_2 = (0, 1/\sqrt{2}, -1\sqrt{2}) \quad \text{and} \quad \mathbf{u}_3 = (4/\sqrt{18}, -1\sqrt{18}, -1\sqrt{18}).$$

Thus

$$P = \begin{bmatrix} 1/3 & 2/3 & 2/3 \\ 0 & 1/\sqrt{2} & -1/\sqrt{2} \\ 4/3\sqrt{2} & -1/3\sqrt{2} & -1/3\sqrt{2} \end{bmatrix}.$$

We emphasize that the above matrix P is not unique.

350. In order for $\mathbf{v}'$ to be orthogonal to $\mathbf{w}$ we must have $\langle \mathbf{v} - c\mathbf{w}, \mathbf{w} \rangle = 0$, or $\langle \mathbf{v}, \mathbf{w} \rangle - c\langle \mathbf{w}, \mathbf{w} \rangle = 0$, or $\langle \mathbf{v}, \mathbf{w} \rangle = c\langle \mathbf{w}, \mathbf{w} \rangle$. Thus $c = \langle \mathbf{v}, \mathbf{w} \rangle / \langle \mathbf{w}, \mathbf{w} \rangle$. Conversely, suppose $c = \langle \mathbf{v}, \mathbf{w} \rangle / \langle \mathbf{w}, \mathbf{w} \rangle$. Then

$$\langle \mathbf{v} - c\mathbf{w}, \mathbf{w} \rangle = \langle \mathbf{v}, \mathbf{w} \rangle - c\langle \mathbf{w}, \mathbf{w} \rangle = \langle \mathbf{v}, \mathbf{w} \rangle - \frac{\langle \mathbf{v}, \mathbf{w} \rangle}{\langle \mathbf{w}, \mathbf{w} \rangle} \langle \mathbf{w}, \mathbf{w} \rangle = 0.$$

351. Compute $\langle \mathbf{v}, \mathbf{w} \rangle = 0 - 1 + 2 = 1$ and $\|\mathbf{w}\|^2 = 0 + 1 + 1 = 2$. Hence $c = \frac{1}{2}$ and $c\mathbf{w} = (0, \frac{1}{2}, \frac{1}{2})$ is the projection of $\mathbf{v}$ along $\mathbf{w}$.

352. First find an orthogonal basis of U using the Gram–Schmidt algorithm. First set $\mathbf{w}_1 = \mathbf{u}_1 = (1, 1, 1, 1)$. Next find

$$\mathbf{v}_2 - \frac{\langle \mathbf{v}_2, \mathbf{w}_1 \rangle}{\|\mathbf{w}_1\|^2} \mathbf{w}_1 = (1, 2, 4, 5) - \frac{12}{4}(1, 1, 1, 1) = (-2, -1, 1, 2).$$

Set $\mathbf{w}_2 = (-2, -1, 1, 2)$. Then find

$$\mathbf{v}_3 - \frac{\langle \mathbf{v}_3, \mathbf{w}_1 \rangle}{\|\mathbf{w}_1\|^2} \mathbf{w}_1 - \frac{\langle \mathbf{v}_3, \mathbf{w}_2 \rangle}{\|\mathbf{w}_2\|^2} \mathbf{w}_2 = (1, -3, -4, -2) - \frac{-8}{4}(1, 1, 1, 1) - \frac{-7}{10}(-2, -1, 1, 2) =$$
$$\left(\frac{8}{5}, -\frac{17}{10}, \frac{13}{10}, \frac{7}{5} \right).$$

Clear fractions to obtain $\mathbf{w}_3 = (16, -17, -13, 14)$. (In hand calculations, it is usually simpler to clear fractions, as this does not affect the orthogonality.) Last, normalize the orthogonal basis $\mathbf{w}_1 = (1, 1, 1, 1)$, $\mathbf{w}_2 = (-2, -1, 1, 2)$, $\mathbf{w}_3 = (16, -17, -13, 14)$. Since $\|\mathbf{w}_1\|^2 = 4$, $\|\mathbf{w}_2\|^2 = 10$, $\|\mathbf{w}_3\|^2 = 910$, the following vectors form an orthonormal basis of U:

$$\mathbf{u}_1 = \frac{1}{2}(1, 1, 1, 1), \qquad \mathbf{u}_2 = \frac{1}{\sqrt{10}}(-2, -1, 1, 2), \qquad \mathbf{u}_3 = \frac{1}{\sqrt{910}}(16, -17, -13, 14).$$

353. First set $\mathbf{w}_1 = \mathbf{v}_1 = (1, 1, 1)$. Then find

$$\mathbf{v}_2 - \frac{\langle \mathbf{v}_2, \mathbf{w}_1 \rangle}{\|\mathbf{w}_1\|^2} \mathbf{w}_1 = (0, 1, 1) - \frac{2}{3}(1, 1, 1) = \left(-\frac{2}{3}, \frac{1}{3}, \frac{1}{3} \right).$$

Clear fractions to obtain $\mathbf{w}_2 = (-2, 1, 1)$. Next find

$$\mathbf{v}_3 - \frac{\langle \mathbf{v}_3, w_1 \rangle}{\|\mathbf{w}_1\|^2} - \frac{\langle \mathbf{v}_3, \mathbf{w}_2 \rangle}{\|\mathbf{w}_2\|^2} = (0, 0, 1) - \frac{1}{3}(1, 1, 1) - \frac{1}{6}(-2, 1, 1) = \left(0, -\frac{1}{2}, \frac{1}{2} \right).$$

Clear fractions to obtain $\mathbf{w}_3 = (0, -1, 1)$. Normalize $\{\mathbf{w}_1, \mathbf{w}_2, \mathbf{w}_3\}$ to obtain the following required orthonormal basis of $\mathbf{R}^3$:

$$\left\{ \mathbf{u}_1 = \left(\frac{1}{\sqrt{3}}, \frac{1}{\sqrt{3}}, \frac{1}{\sqrt{3}} \right), \mathbf{u}_2 = \left(-\frac{2}{\sqrt{6}}, \frac{1}{\sqrt{6}}, \frac{1}{\sqrt{6}} \right), \mathbf{u}_3 = \left(0, -\frac{1}{\sqrt{2}}, \frac{1}{\sqrt{2}} \right) \right\}.$$

354. Note that $c_i = \langle \mathbf{v}, \mathbf{u}_i \rangle$ since $\|\mathbf{u}_i\| = 1$. Using $\langle \mathbf{u}_i, \mathbf{u}_j \rangle = 0$ for $i \neq j$, we get

$$0 \leq \langle \mathbf{v} - \sum c_k \mathbf{u}_k, \mathbf{v} - \sum c_k, \mathbf{u}_k \rangle = \langle \mathbf{v}, \mathbf{v} \rangle - 2\langle \mathbf{v}, \sum c_k \mathbf{u}_k \rangle + \sum c_k^2 = \langle \mathbf{v}, \mathbf{v} \rangle - \sum 2c_k \langle \mathbf{v}, \mathbf{v}_k \rangle +$$
$$\sum ck_k^2 = \langle \mathbf{v}, \mathbf{v} \rangle - \sum 2c_k^2 + \sum c_k^2 = \langle \mathbf{v}, \mathbf{v} \rangle - \sum c_k^2.$$

This gives us our inequality.

355. Compute $\langle \mathbf{u}_1, \mathbf{u}_1 \rangle = 1 + 1 + 0 = 2$, $\langle \mathbf{u}_1, \mathbf{u}_2 \rangle = 1 + 2 + 0 = 3$, $\langle \mathbf{u}_1, \mathbf{u}_3 \rangle = 1 + 3 + 0 = 4$, $\langle \mathbf{u}_2, \mathbf{u}_2 \rangle = 1 + 4 + 9 = 14$, $\langle \mathbf{u}_2, \mathbf{u}_3 \rangle = 1 + 6 + 15 = 22$, $\langle \mathbf{u}_3, \mathbf{u}_3 \rangle = 1 + 9 + 25 = 35$.
   Thus

$$A = \begin{bmatrix} 2 & 3 & 4 \\ 3 & 14 & 22 \\ 4 & 22 & 35 \end{bmatrix}.$$

356. We have $\langle \mathbf{e}_1, \mathbf{e}_1 \rangle = 1$, $\langle \mathbf{e}_1, \mathbf{e}_2 \rangle = 0$, $\langle \mathbf{e}_1, \mathbf{e}_3 \rangle = 0$, $\langle \mathbf{e}_2, \mathbf{e}_2 \rangle = 1$, $\langle \mathbf{e}_2, \mathbf{e}_3 \rangle = 0$, $\langle \mathbf{e}_3, \mathbf{e}_3 \rangle = 1$. Thus the identity matrix I represents the usual inner product on $\mathbf{R}^3$ with respect to the usual basis E of $\mathbf{R}^3$.

357. $\langle \mathbf{f}, \mathbf{g} \rangle = \displaystyle\int_{-1}^{1} (t + 2)(t^2 - 3t + 4)\, dt = \int_{-1}^{1} (t^3 - t^2 - 2t + 8)\, dt = \left[\dfrac{t^4}{4} - \dfrac{t^3}{3} - t^2 + 8t \right]_{-1}^{1} = \dfrac{46}{3}$.

358. Here we use the fact that if $r + s = n$ then

$$\langle t^r, t^s \rangle = \int_{-1}^{1} t^n \, dt = \left[\frac{t^{n+1}}{n+1} \right]_{-1}^{1} = \begin{cases} 2/(n+1) & \text{if } n \text{ is even} \\ 0 & \text{if } n \text{ is odd.} \end{cases}$$

Then $\langle 1, 1 \rangle = 2$, $\langle 1, t \rangle = 0$, $\langle 1, t^2 \rangle = \frac{2}{3}$, $\langle t, t \rangle = \frac{2}{3}$, $\langle t, t^2 \rangle = 0$, $\langle t^2, t^2 \rangle = \frac{2}{5}$. Thus

$$A = \begin{bmatrix} 2 & 0 & \frac{2}{3} \\ 0 & \frac{2}{3} & 0 \\ \frac{2}{3} & 0 & \frac{2}{3} \end{bmatrix}.$$

359. We have $[f]^{\mathrm{T}} = (2, 1, 0)$ and $[g]^{\mathrm{T}} = (4, -3, 1)$ relative to the given basis. Then

$$[\mathbf{f}]^{\mathrm{T}} A[\mathbf{g}] = (2, 1, 0) \begin{bmatrix} 2 & 0 & \frac{2}{3} \\ 0 & \frac{2}{3} & 0 \\ \frac{2}{3} & 0 & \frac{2}{3} \end{bmatrix} \begin{bmatrix} 4 \\ -3 \\ 1 \end{bmatrix} = \left(4, \frac{2}{3}, \frac{4}{3} \right) \begin{bmatrix} 4 \\ -3 \\ 1 \end{bmatrix} = \frac{46}{3} = \langle \mathbf{f}, \mathbf{g} \rangle.$$

360. A is symmetric since $\langle \mathbf{e}_i, \mathbf{e}_j \rangle = \langle \mathbf{e}_j, \mathbf{e}_i \rangle$. Let $\mathbf{x}$ be any nonzero vector in $\mathbf{R}^n$. Then $[\mathbf{u}] = \mathbf{x}$ for some nonzero vector $\mathbf{u} \in V$. we have $\mathbf{x}^{\mathrm{T}} A\mathbf{x} = [\mathbf{u}]^{\mathrm{T}} A[\mathbf{u}] = \langle \mathbf{u}, \mathbf{u} \rangle > 0$. Thus A is positive definite.

361. $\langle \mathbf{u}, k\mathbf{v} \rangle = \overline{\langle k\mathbf{v}, \mathbf{u} \rangle} = \overline{k \langle \mathbf{v}, \mathbf{u} \rangle} = \bar{k}\, \overline{\langle \mathbf{v}, \mathbf{u} \rangle} = \bar{k}\, \overline{\langle \mathbf{u}, \mathbf{v} \rangle} = \bar{k}\, \langle \mathbf{u}, \mathbf{v} \rangle$

362. **(A)** $\langle (2 - 4i)\mathbf{u}, \mathbf{v} \rangle = (2 - 4i)\langle \mathbf{u}, \mathbf{v} \rangle = (2 - 4i)(3 + 2i) = 14 - 4i$
   (B) $\langle \mathbf{u}, (4 + 3i)\mathbf{v} \rangle = \overline{(4 + 3i)}\langle \mathbf{u}, \mathbf{v} \rangle = (4 - 3i)(3 + 2i) = 18 - i$

363. $\langle (3 - 6i)\mathbf{u}, (5 - 2i)\mathbf{v} \rangle = (3 - 6i)\overline{(5 - 2i)}\langle \mathbf{u}, \mathbf{v} \rangle = (3 - 6i)(5 + 2i)(3 + 2i) = 137 - 30i$

364. **(A)** Recall that the conjugate of the second vector appears in the inner product:
$\langle \mathbf{u}, \mathbf{v} \rangle = (1-i)\overline{(2-5i)} + (2+3i)\overline{(3-i)} = (1-i)(2+5i) + (2+3i)(3+i) = 7+3i+3+11i = 10+14i$.

(B) $\langle \mathbf{v}, \mathbf{u} \rangle = (2-5i)\overline{(1-i)} + (3-i)\overline{(2+3i)} = (2-5i)(1+i)(3-i)(2-3i) = 7-3i+3-11i = 10 - 14i$. [As expected from [CIP$_2$], $\langle \mathbf{v}, \mathbf{u} \rangle = \overline{\langle \mathbf{u}, \mathbf{v} \rangle}$.]

365. **(A)** Recall that $z\bar{z} = a^2 + b^2$ when $z = a+bi$. Use $\|\mathbf{u}\|^2 = \langle \mathbf{u}, \mathbf{u} \rangle = z_1\bar{z}_1 + z_2\bar{z}_2 + \cdots + z_n\bar{z}_n$ where $\mathbf{u} = (z_1, z_2, \ldots, z_n)$. Compute $\|\mathbf{u}\|^2 = 1^2 + (-1)^2 + 2^2 + 3^2 = 1+1+2+9 = 13$, or $\|\mathbf{u}\| = \sqrt{13}$.

(B) $\|\mathbf{v}\|^2 = 4+25+9+1 = 39$ and so $\|\mathbf{v}\| = \sqrt{39}$.

Chapter 10: Eigenvalues, Eigenvectors, Diagonalization

366. **(A)** Here $\mathrm{tr}(A) = 1 + 2 = 3$ and $|A| = 2 - 6 = -4$. Hence $\Delta(t) = t^2 - \mathrm{tr}(A) + |A| = t^2 - 3t - 4$.

(B) $\Delta(A) = A^2 - 3A - 4I = \begin{bmatrix} 7 & 6 \\ 9 & 10 \end{bmatrix} + \begin{bmatrix} -3 & -6 \\ -9 & -6 \end{bmatrix} + \begin{bmatrix} -4 & 0 \\ 0 & -4 \end{bmatrix} = \begin{bmatrix} 0 & 0 \\ 0 & 0 \end{bmatrix}$.

367. **(A)** Here $\mathrm{tr}(A) = -2 + 9 = 7$ and $\det(A) = -18 + 24 = 6$. Hence $\Delta(t) = t^2 - 7t + 6$.

(B) Here $\mathrm{tr}(B) = 4 + (-7) = -3$ and $\det(B) = -28 + 15 = -13$. Hence $\Delta(t) = t^2 + 3t - 13$. (We emphasize that it is the negative of the trace which is the coefficient of t^{n-1}.)

368. Here $\mathrm{tr}(A) = 1 + 4 + 2 = 7$. The cofactors of the diagonal elements follow:

$$A_{11} = \begin{vmatrix} 4 & 1 \\ 7 & 2 \end{vmatrix} = 1, \quad A_{22} = \begin{vmatrix} 1 & 3 \\ 2 & 2 \end{vmatrix} = -4, \quad A_{33} = \begin{vmatrix} 1 & 2 \\ 5 & 4 \end{vmatrix} = -6.$$

Thus $A_{11} + A_{22} + A_{33} = -9$, and $|A| = 8 + 4 + 105 - 24 - 7 - 20 = 66$. Therefore $\Delta(A) = t^3 - 7t^2 - 9t - 66$.

369. Here $\mathrm{tr}(B) = 1 + 3 + 9 = 13$. The cofactors of the diagonal elements follow:

$$B_{11} = \begin{vmatrix} 3 & 2 \\ 3 & 9 \end{vmatrix} = 21, \quad B_{22} = \begin{vmatrix} 1 & 2 \\ 1 & 9 \end{vmatrix} = 7, \quad B_{33} = \begin{vmatrix} 1 & 1 \\ 0 & 3 \end{vmatrix} = 3.$$

Thus $B_{11} + B_{22} + B_{33} = 31$ and $|B| = 27 + 2 + 0 - 6 - 6 - 0 = 17$. Hence $\Delta(B) = t^3 - 13t^2 + 31t - 17$.

370. Since R is triangular, $\Delta(t) = (t-1)(t-2)(t-3)(t-4)$.

371. Note that S is block triangular with diagonal blocks $A_1 = \begin{bmatrix} 2 & 5 \\ 1 & 4 \end{bmatrix}$ and $A_2 = \begin{bmatrix} 6 & -5 \\ 2 & 3 \end{bmatrix}$. Thus $\Delta(t) = \Delta_{A_1}(t)\Delta_{A_2}(t) = (t^2 - 6t + 3)(t^2 - 9t + 28)$.

372. Note that T is block triangular with diagonal blocks $[5]$, $\begin{bmatrix} 3 & 6 \\ -3 & 5 \end{bmatrix}$, and $[7]$. Thus

$\Delta(t) = (t-5)(t^2 - 8t + 33)(t-7)$.

373. (A) $A\mathbf{v}_1 = \begin{bmatrix} 1 & 2 \\ 3 & 2 \end{bmatrix}\begin{bmatrix} 2 \\ 3 \end{bmatrix} = \begin{bmatrix} 8 \\ 12 \end{bmatrix} = 4\begin{bmatrix} 2 \\ 3 \end{bmatrix} = 4\mathbf{v}_1$.

Thus $\mathbf{v}_1$ is an eigenvector of A belonging to $\lambda_1 = 4$.

(B) $A\mathbf{v}_2 = \begin{bmatrix} 1 & 2 \\ 3 & 2 \end{bmatrix}\begin{bmatrix} 1 \\ -1 \end{bmatrix} = \begin{bmatrix} -1 \\ 1 \end{bmatrix} = (-1)\mathbf{v}_2$.

Thus $\mathbf{v}_2$ is an eigenvector of A belonging to $\lambda_2 = -1$.

374. By Question 373, A has two linearly independent eigenvectors $\begin{bmatrix} 2 \\ 3 \end{bmatrix}$ and $\begin{bmatrix} 1 \\ -1 \end{bmatrix}$. Set $P = \begin{bmatrix} 2 & 1 \\ 3 & -1 \end{bmatrix}$, and so $P^{-1} = \begin{bmatrix} \frac{1}{5} & \frac{1}{5} \\ \frac{3}{5} & -\frac{2}{2} \end{bmatrix}$. Then A is similar to the diagonal matrix

$$B = P^{-1} AP = \begin{bmatrix} \frac{1}{5} & \frac{1}{5} \\ \frac{3}{5} & -\frac{2}{5} \end{bmatrix}\begin{bmatrix} 1 & 2 \\ 3 & 2 \end{bmatrix}\begin{bmatrix} 2 & 1 \\ 3 & -1 \end{bmatrix} = \begin{bmatrix} 4 & 0 \\ 0 & -1 \end{bmatrix}.$$

As expected, the diagonal elements 4 and -1 of the diagonal matrix B are the eigenvalues corresponding to the given eigenvectors.

375. (A) Here $\Delta(t) = t^2 - \text{tr}(A)t + |A| = t^2 - 5t + 4 = (t-1)(t-4)$. Hence $\lambda_1 = 1$ and $\lambda_2 = 4$ are eigenvalues of A. We find corresponding eigenvectors:

(1) Subtract $\lambda_1 = 1$ down the diagonal of A to obtain $M = \begin{bmatrix} 1 & 2 \\ 1 & 2 \end{bmatrix}$, which corresponds to the homogeneous system $x + 2y = 0$. Here $\mathbf{v}_1 = (2, -1)$ is a nonzero solution of the system and so is an eigenvector of A belonging to $\lambda_1 = 1$.

(2) Subtract $\lambda_2 = 4$ down the diagonal of A to obtain $M = \begin{bmatrix} -2 & 2 \\ 1 & -1 \end{bmatrix}$, which corresponds to the homogeneous system $x - y = 0$. Here $\mathbf{v}_2 = (1, 1)$ is a nonzero solution and so is an eigenvector of A belonging to $\lambda_2 = 4$.

Then $S = \{\mathbf{v}_1 = (2, -1), \mathbf{v}_2 = (1, -1)\}$.

(B) Yes, since A has two independent eigenvectors. Let P be the matrix whose columns are $\mathbf{v}_1$ and $\mathbf{v}_2$. Then $P = \begin{bmatrix} 2 & 1 \\ -1 & -1 \end{bmatrix}$ and $D = P^{-1}AP = \begin{bmatrix} 1 & 0 \\ 0 & 4 \end{bmatrix}$.

376. (A) Here $\Delta(t) = |tI - A| = t^2 + 1$. Since $t^2 + 1$ has no solution in $\mathbf{R}$, A has no eigenvalues and hence no eigenvectors.

(B) Viewed as a real matrix, A has no eigenvectors, and hence A is not diagonalizable.

377. (A) Again $\Delta(t) = |tI - A| = t^2 + 1$. Now, however, $\lambda_1 = i$ and $\lambda_2 = -i$ are eigenvalues of A.

(1) Substitute $t = i$ in $tI - B$ to obtain the homogeneous system

$$\begin{bmatrix} i-1 & 1 \\ -2 & i+1 \end{bmatrix}\begin{bmatrix} x \\ y \end{bmatrix} = \begin{bmatrix} 0 \\ 0 \end{bmatrix} \quad \text{or} \quad \begin{cases} (i-1)x + y = 0 \\ -2x + (i+1)y = 0 \end{cases} \quad \text{or} \quad (i-1)x + y = 0.$$

The system has only one independent solution, $x = 1$, $y = 1 - i$. Thus $\mathbf{v}_1 = (1, 1 - i)$ is an eigenvector that spans the eigenspace of $\lambda_1 = i$.

(2) Substitute $t = -i$ into $tI - B$ to obtain the homogeneous system

$$\begin{bmatrix} -i-1 & 1 \\ -2 & -i-1 \end{bmatrix}\begin{bmatrix} x \\ y \end{bmatrix} = \begin{bmatrix} 0 \\ 0 \end{bmatrix} \quad \text{or} \quad \begin{cases} (-i-1)x + y = 0 \\ -2x + (-i-1)y = 0 \end{cases} \quad \text{or} \quad (-i-1)x + y = 0.$$

The system has only one independent solution, $x = 1$, $y = 1 + i$. Thus $\mathbf{v}_2 = (1, 1 + i)$ is an eigenvector of A that spans the eigenspace of $\lambda_2 = -i$.
Then $S = \{\mathbf{v}_1 = (1, 1 - i), \mathbf{v}_2 = (1, 1 + i)\}$.

(B) As a complex matrix, A is diagonalizable. Let P be the matrix whose columns are $\mathbf{v}_1$ and $\mathbf{v}_2$; that is, $P = \begin{bmatrix} 1 & 1 \\ 1-i & 1+i \end{bmatrix}$. Then $P^{-1}AP = \begin{bmatrix} i & 0 \\ 0 & -i \end{bmatrix}$.

378. (A) Here $\Delta(t) = t^2 - \text{tr}(B)t + |B| = t^2 - 3t - 10 = (t - 5)(t + 2)$. Thus $\lambda_1 = 5$ and $\lambda_2 = -2$ are the eigenvalues of B.

(1) Subtract $\lambda_1 = 5$ down the diagonal of B to obtain $M = \begin{bmatrix} -3 & 4 \\ 3 & -4 \end{bmatrix}$, which corresponds to the homogeneous system $3x - 4y = 0$. Here $\mathbf{v}_1 = (4, 3)$ is a nonzero solution.

(2) Subtract $\lambda_2 = -2$ (or add 2) down the diagonal of B to obtain $M = \begin{bmatrix} 4 & 4 \\ 3 & 3 \end{bmatrix}$, which corresponds to the system $x + y = 0$, which has a nonzero solution $\mathbf{v}_2 = (1, -1)$.
Then $S = \{\mathbf{v}_1 = (4, 3), \mathbf{v}_2 = (1, -1)\}$.

(B) Yes, since B has two independent eigenvectors. Let P be the matrix whose columns are $\mathbf{v}_1$ and $\mathbf{v}_2$. Then $P = \begin{bmatrix} 4 & 1 \\ 3 & -1 \end{bmatrix}$ and $D = P^{-1}AP = \begin{bmatrix} 5 & 0 \\ 0 & -2 \end{bmatrix}$.

379. (A) Here $\text{tr}(C) = 4 + 5 + 2 = 11$. The cofactors of the diagonal elements follow: $C_{11} = 12$, $C_{22} = 9$, $C_{33} = 18$. Thus $\Sigma C_{ii} = 39$ and $|C| = 40 - 2 - 2 + 5 + 8 - 4 = 45$. Hence $\Delta(C) = t^3 - 11t^2 - 39t - 45$.

(B) Assuming $\Delta(t)$ has a rational root, it must be among ± 1, ± 3, ± 5, ± 9, ± 45. Testing, we get

$$3 \;\underline{\big|\; \begin{array}{rrr} 1-11 & +39 & -45 \\ 3 & -24 & +45 \end{array}}$$
$$\begin{array}{rrr} 1-8 & +15 & +\,0. \end{array}$$

Thus $t = 3$ is a root of $\Delta(t)$ and $\Delta(t) = (t - 3)(t^2 - 8t + 15) = (t - 3)^2(t - 5)$. Accordingly, $\lambda_1 = 3$ and $\lambda_2 = 5$ are the eigenvalues of C.

380. Find independent eigenvectors for each eigenvalue of C.

    (1) Subtract $\lambda_1 = 3$ down the diagonal of C to obtain the homogeneous system $x + y - z = 0$. Here $\mathbf{u} = (1, -1, 0)$ and $\mathbf{v} = (1, 0, 1)$ are two independent solutions.

    (2) Subtract $\lambda_2 = 5$ down the diagonal of C to obtain the homogeneous system

$$-x + y - z = 0, \quad 2x - 2z = 0, \quad x + y - 3z = 0.$$

The system yields a nonzero solution $\mathbf{w} = (1, 2, 1)$.

    Thus $S = \{\mathbf{u}, \mathbf{v}, \mathbf{w}\}$. (We emphasize that S is not unique.)

381. Yes, since C has three independent eigenvectors. Let P be the matrix whose columns are $\mathbf{u}, \mathbf{v}, \mathbf{w}$. Then $P^{-1} CP$ is the diagonal matrix whose diagonal entries are the corresponding eigenvalues. That is, $P = \begin{bmatrix} 1 & 1 & 1 \\ -1 & 0 & 2 \\ 0 & 1 & 1 \end{bmatrix}$ and $D = P^{-1}AP = \begin{bmatrix} 3 & & \\ & 3 & \\ & & 5 \end{bmatrix}$.

382. Here $P^{-1} AP = D$ where $P = \begin{bmatrix} 1 & 1 \\ 3 & 4 \end{bmatrix}$ and $D = \begin{bmatrix} 2 & 0 \\ 0 & 3 \end{bmatrix}$. Compute $P^{-1} = \begin{bmatrix} 4 & -1 \\ -3 & 1 \end{bmatrix}$.

Then $A = PDP^{-1} = \begin{bmatrix} 1 & 1 \\ 3 & 4 \end{bmatrix}\begin{bmatrix} 2 & 0 \\ 0 & 3 \end{bmatrix}\begin{bmatrix} 4 & -1 \\ -3 & 1 \end{bmatrix} = \begin{bmatrix} -1 & 1 \\ -12 & 6 \end{bmatrix}$.

383. Here $P^{-1}BP = D$ where $P = \begin{bmatrix} 1 & 3 \\ 2 & 5 \end{bmatrix}$ and $D = \begin{bmatrix} -1 & 0 \\ 0 & 2 \end{bmatrix}$. Compute $P^{-1} = \begin{bmatrix} -5 & 3 \\ 2 & -1 \end{bmatrix}$.

Then $B = PDP^{-1} = \begin{bmatrix} 1 & 3 \\ 2 & 5 \end{bmatrix}\begin{bmatrix} -1 & 0 \\ 0 & 2 \end{bmatrix}\begin{bmatrix} -5 & 3 \\ 2 & -1 \end{bmatrix} = \begin{bmatrix} 17 & -9 \\ 30 & -16 \end{bmatrix}$.

384. Here $P^{-1}CP = D$ where $P = \begin{bmatrix} 1 & 1 & 1 \\ 0 & 1 & 2 \\ 1 & 2 & 4 \end{bmatrix}$ and $D = \begin{bmatrix} 1 & 0 & 0 \\ 0 & 2 & 0 \\ 0 & 0 & 3 \end{bmatrix}$. Compute

$P^{-1} = \begin{bmatrix} 0 & -2 & 1 \\ 2 & 3 & -2 \\ -1 & -1 & 1 \end{bmatrix}$.

Then $C = PDP^{-1} = \begin{bmatrix} 1 & 1 & 1 \\ 0 & 1 & 2 \\ 1 & 2 & 4 \end{bmatrix}\begin{bmatrix} 1 & 0 & 0 \\ 0 & 2 & 0 \\ 0 & 0 & 3 \end{bmatrix}\begin{bmatrix} 0 & -2 & 1 \\ 2 & 3 & -2 \\ -1 & -1 & 1 \end{bmatrix} = \begin{bmatrix} 1 & 1 & 0 \\ -2 & 0 & 2 \\ -4 & -2 & 5 \end{bmatrix}$.

385. Here $\Delta(t) = t^2 - \operatorname{tr}(A) + |A| = t^2 - 6t + 5 = (t - 1)(t - 5)$. Thus $\lambda = 1$ and $\lambda = 5$ are the eigenvalues of A.

386. Find eigenvectors for each eigenvalue of A.

(1) Subtract $\lambda = 1$ down the diagonal of A to obtain the corresponding homogeneous system $2x + y = 0$. A nonzero solution is $\mathbf{u}_1 = (1, -1)$.

(2) Subtract $\lambda = 5$ down the diagonal of A to obtain the corresponding homogeneous system $\begin{matrix} -2x + 2y = 0 \\ 2x - 2y = 0 \end{matrix}$ or, simply, $x - y = 0$. A nonzero solution is $\mathbf{u}_2 = (1, 1)$.

Thus $S = \{\mathbf{u}_1 = (1, -1), \mathbf{u}_2 = (1, 1)\}$. (As expected, $\mathbf{u}_1$ and $\mathbf{u}_2$ are orthogonal.)

387. Normalize $\mathbf{u}_1$ and $\mathbf{u}_2$ to obtain the unit vectors $\hat{\mathbf{u}}_1 = (1/\sqrt{2}, -1/\sqrt{2})$ and $\hat{\mathbf{u}}_2 = (1/\sqrt{2}, 1/\sqrt{2})$.

Let P be the matrix whose columns are $\hat{\mathbf{u}}_1$ and $\hat{\mathbf{u}}_2$. Then $P = \begin{bmatrix} 1/\sqrt{2} & 1/\sqrt{2} \\ -1/\sqrt{2}) & 1/\sqrt{2} \end{bmatrix}$ and $D = P^{-1}AP = \begin{bmatrix} 1 & 0 \\ 0 & 5 \end{bmatrix}$.

388. Here $\Delta(t) = t^2 - \text{tr}(B) + |B| = t^2 - 6t - 16 = (t - 8)(t + 2)$. Thus $\lambda = 8$ and $\lambda = -2$ are the eigenvalues of A.

389. (1) Subtract $\lambda = 8$ down the diagonal of A to obtain the corresponding homogeneous system $-x + 3y = 0$. A nonzero solution is $\mathbf{u}_1 = (3, 1)$.

(2) Subtract $\lambda = -2$ (or add 2) down the diagonal of A to obtain the corresponding homogeneous system $3x + y = 0$. A nonzero solution is $\mathbf{u}_2 = (1, -3)$.

Thus $S = \{\mathbf{u}_1 = (3, 1), \mathbf{u}_2 = (1, -3)\}$. (As expected, $\mathbf{u}_1$ and $\mathbf{u}_2$ are orthogonal.)

390. Normalize $\mathbf{u}_1$ and $\mathbf{u}_2$ to obtain the unit vectors $\hat{\mathbf{u}}_1 = (3/\sqrt{10}, -1/\sqrt{10})$ and $\hat{\mathbf{u}}_2 = (1/\sqrt{10}, -3/\sqrt{10})$. Let P be the matrix whose columns are $\hat{\mathbf{u}}_1$ and $\hat{\mathbf{u}}_2$. Then

$$P = \begin{bmatrix} 3/\sqrt{10} & 1/\sqrt{10} \\ -1/\sqrt{10}) & -3/\sqrt{10} \end{bmatrix} \quad \text{and} \quad D = P^{-1}AP = \begin{bmatrix} 8 & 0 \\ 0 & -2 \end{bmatrix}.$$

391. Here $\text{tr}(C) = 14$. The cofactors of the diagonal elements are $C_{11} = 11$, $C_{22} = 3$, $C_{33} = 11$. Thus $\Sigma C_{ii} = 25$ and $|C| = 40 + 9 + 3 - 10 - 18 - 18 = 12$. Hence $\Delta(t) = t^3 - 14t^2 + 25t - 12$. Assuming $\Delta(t)$ has a rational root, it must divide 12. Testing, we get $t = 1$ as a root. Hence $t - 1$ divides $\Delta(t)$. We get $\Delta(t) = (t - 1)(t^2 - 13t + 12) = (t - 1)^2 (t - 12)$. Thus $\lambda_1 = 1$ (multiplicity two) and $\lambda_2 = 12$ are the eigenvalues of C.

392. Find orthogonal eigenvectors for each eigenvalue of C.

(1) Subtract $\lambda_1 = 1$ down the diagonal of C to obtain the homogeneous system

$$x + 3y + z = 0, \quad 3x + 9y + 3z = 0, \quad x + 3y + z = 0$$

or, simply, $x + 3y + z = 0$. One solution is $\mathbf{u} = (0, 1, -3)$. We seek another solution $\mathbf{v} = (a, b, c)$ which is also orthogonal to $\mathbf{u}$. This gives us the system:

$$a + 3b + c = 0$$
$$b - 3c = 0.$$

A solution is $\mathbf{v} = (10, -3, -1)$.

(2)   Subtract $\lambda_2 = 12$ down the diagonal of C to obtain the homogeneous system
$-10x + 3y + z = 0, \quad 3x - 2y + 3z = 0, \quad x + 3y - 10z = 0.$

The system yields a nonzero solution $\mathbf{w} = (1, 3, 1)$.
   Thus $S = \{\mathbf{u}, \mathbf{v}, \mathbf{w}\}$. (We emphasize that S is not unique.)

393. Normalize $\mathbf{u}$, $\mathbf{v}$, $\mathbf{w}$ to obtain $\mathbf{u} = (0, 1/\sqrt{10}, -3/\sqrt{10})$, $\mathbf{v} = (10/\sqrt{110}, -3/\sqrt{110},$
$-1/\sqrt{110})$, $\mathbf{w} = (1/\sqrt{11}, 3/\sqrt{11}, 1/\sqrt{11})$. Let P be the matrix whose columns are $\hat{\mathbf{u}}$, $\hat{\mathbf{v}}$, $\hat{\mathbf{w}}$. Then
$P^{-1} CP$ is the diagonal matrix whose diagonal entries are the corresponding eigenvalues.
That is, $P = [\hat{\mathbf{u}}, \hat{\mathbf{v}}, \hat{\mathbf{w}}]$ and $D = P^{-1}AP = \text{diag}(1, 1, 12)$.

394. $q(x, y) = (x, y) \begin{bmatrix} 5 & -3 \\ -3 & 8 \end{bmatrix} \begin{bmatrix} x \\ y \end{bmatrix} = (5x - 3y, -3x + 8y) \begin{bmatrix} x \\ y \end{bmatrix}$
$\qquad = 5x^2 - 3xy - 3xy + 8y^2 = 5x^2 - 6xy + 8y^2.$

395. Here $q(x, y, z) = 3x^2 - 4y^2 + 6z^2$. (There are no cross-product terms.)

396. $q(x, y, z) = 2x^2 - 10xy - 6y^2 + 2xz - 14yz + 9z^2$. (As usual, we assume that x, y, z
are the first, second, and third variables, respectively.)

397. The symmetric matrix $A = [a_{ij}]$ representing $q(x_1, \ldots, x_n)$ has the diagonal entry a_{ij}
equal to the coefficient of x_i^2 and has the entries a_{ij} and a_{ji} each equal to half the coefficient
of $x_i x_j$. Thus

$$A = \begin{bmatrix} 3 & 2 & 4 \\ 2 & -1 & -3 \\ 4 & -3 & 1 \end{bmatrix}.$$

398. Here $B = \begin{bmatrix} 4 & \frac{5}{2} \\ \frac{5}{2} & -7 \end{bmatrix}$. (Division by 2 may introduce fractions even though the coef-

ficients in q are integers.)

399. Even though only x and y appear in the polynomial, the expression $q(x, y, z)$ indicates
that there are three variables. In other words, $q(x, y, z) = 0x^2 + 4xy + 5y^2 + 0xz + 0yz + 0z^2$.
Thus

$$C = \begin{bmatrix} 0 & 2 & 0 \\ 2 & 5 & 0 \\ 0 & 0 & 0 \end{bmatrix}.$$

400. Here $D = \begin{bmatrix} 1 & 0 & 1/2 \\ 0 & 0 & -1 \\ 1/2 & -1 & 0 \end{bmatrix}.$

401. Substitute for x and y in q to obtain

$$q(s, t) = 3(s - 3t)^2 + 2(s - 3t)(2s + t) - (2s + t)^2$$
$$= 3(s^2 - 6st + 9t^2) + 2(2s^2 - 5st - 3t^2) - (s^2 + 4st + t^2) = 3s^2 - 32st + 20t^2.$$

402. We have $A = \begin{bmatrix} 3 & 1 \\ 1 & -1 \end{bmatrix}$ and $q(\mathbf{x}) = \mathbf{x}^T A \mathbf{x}$ where $\mathbf{x} = (x, y)^T$.

403. We have $\begin{bmatrix} x \\ y \end{bmatrix} = \begin{bmatrix} 1 & -3 \\ 2 & 1 \end{bmatrix}\begin{bmatrix} s \\ t \end{bmatrix}$. Thus $P = \begin{bmatrix} 1 & -3 \\ 2 & 1 \end{bmatrix}$ and $\mathbf{x} = P\mathbf{y}$ where $\mathbf{x} = (x, y)^T$ and $\mathbf{y} = (s, t)^T$.

404. We have $q(\mathbf{x}) = \mathbf{x}^T A \mathbf{x}$ and $\mathbf{x} = P\mathbf{y}$. Thus $\mathbf{x}^T = \mathbf{y}^T P^T$. Therefore

$$q(s, t) = q(\mathbf{y}) = \mathbf{y}^T P^T A P \mathbf{y} = (s, t)\begin{bmatrix} 1 & 2 \\ -3 & 1 \end{bmatrix}\begin{bmatrix} 3 & 1 \\ 1 & -1 \end{bmatrix}\begin{bmatrix} 1 & -3 \\ 2 & 1 \end{bmatrix}\begin{bmatrix} s \\ t \end{bmatrix}$$

$$= (s, t)\begin{bmatrix} 3 & -16 \\ -16 & 20 \end{bmatrix}\begin{bmatrix} s \\ t \end{bmatrix} = 3s^2 - 32st + 20t^2.$$

405. First factor out the coefficient 3 of x^2 from the x^2 term and the xy term, and then complete the square inside the parentheses by adding $4y^2$ and then subtracting the corresponding amount $3(4y^2) = 12y^2$ outside the parentheses. This gives

$$q(x, y) = 3x^2 - 12xy + 7y^2 = 3(x^2 - 4xy) + 7y^2$$
$$= 3(x^2 - 4xy + 4y^2) + 7y^2 - 12y^2 = 3(x - 2y)^2 - 5y^2.$$

Let $s = x - 2y$, $t = y$. Then $x = s + 2t$, $y = t$. This linear substitution yields $q(s, t) = 3s^2 - 5t^2$.

406. Here $\text{tr}(A) = 4$. The cofactors of the diagonal elements are $A_{11} = -1$, $A_{22} = 6$, $A_{33} = 0$. Thus $\Sigma A_{ii} = 5$, and $|A| = -36 - 24 - 24 + 18 + 32 + 36 = 2$. Hence

$$\Delta(t) = t^3 - \text{tr}(A) + (A_{11} + A_{22} + A_{33})t - |A| = t^3 - 4t^2 + 5t - 2 = (t - 2)(t - 1)^2.$$

407. The minimal polynomial $m(t)$ must divide $\Delta(t)$. Also, each irreducible factor of $\Delta(t)$ (that is, $t - 2$ and $t - 1$) must also be a factor of $m(t)$. Thus $m(t)$ is exactly only one of the following: $f(t) = (t - 2)(t - 1)$ or $g(t) = (t - 2)(t - 1)^2$. Testing $f(t)$, we have

$$f(A) = (A - 2I)(A - I) = \begin{bmatrix} 2 & -2 & 2 \\ 6 & -5 & 4 \\ 3 & -2 & 1 \end{bmatrix}\begin{bmatrix} 3 & -2 & 2 \\ 6 & -4 & 4 \\ 3 & 2 & 2 \end{bmatrix} = \begin{bmatrix} 0 & 0 & 0 \\ 0 & 0 & 0 \\ 0 & 0 & 0 \end{bmatrix}.$$

Thus $f(t) = m(t) = (t - 2)(t - 1) = t^2 - 3t + 2$ is the minimal polynomial of A.

408. Here $\text{tr}(B) = 4$. The cofactors of the diagonal elements are $B_{11} = -2$, $B_{22} = 11$, $B_{33} = -4$. Thus $\Sigma B_{ii} = 5$, and $|B| = -60 - 24 - 24 + 16 + 54 + 40 = 2$. Hence

$$\Delta(t) = t^3 - \text{tr}(B) + (B_{11} + B_{22} + B_{33})t - |A| = t^3 - 4t^2 + 5t - 2 = (t - 2)(t - 1)^2.$$

409. The minimal polynomial $m(t)$ is exactly one of the following: $f(t) = (t-2)(t-1)$ or $g(t) = (t-2)(t-1)^2$. Testing $f(t)$, we have

$$f(B) = (B-2I)(B-I) = \begin{bmatrix} 1 & -2 & 2 \\ 4 & -6 & 6 \\ 2 & -3 & 3 \end{bmatrix}\begin{bmatrix} 2 & -2 & 2 \\ 4 & -5 & 6 \\ 2 & -3 & 4 \end{bmatrix} = \begin{bmatrix} -2 & 2 & -2 \\ -4 & 4 & -4 \\ -2 & 2 & -2 \end{bmatrix} \neq 0.$$

Thus $f(t) \neq m(t)$. Accordingly, $m(t) = g(t) = (t-2)(t-1)^2$ is the minimal polynomial of B. [We do not need to compute $g(B)$; we know $g(B) = 0$ by the Cayley–Hamilton theorem.]

410. Note that A is a block diagonal matrix with diagonal blocks

$$A_1 = \begin{bmatrix} 2 & 5 \\ 0 & 2 \end{bmatrix}, \quad A_2 = \begin{bmatrix} 4 & 2 \\ 3 & 5 \end{bmatrix}, \quad A_3 = [7].$$

Then $\Delta(t)$ is the product of the characteristic polynomials $\Delta_1(t)$, $\Delta_2(t)$, and $\Delta_3(t)$ of A_1, A_2, and A_3, respectively. Since A_1 and A_3 are triangular, $\Delta_1(t) = (t-2)^2$ and $\Delta_3(t) = (t-7)$. Also, $\Delta_2(t) = t^2 - 9t + 14 = (t-2)(t-7)$. Thus $\Delta(t) = (t-2)^3(t-7)^2$. [As expected, deg $(m(t)) = 5$.]

411. Note that the minimal polynomials $m_1(t)$, $m_2(t)$, and $m_3(t)$ of the diagonal blocks A_1, A_2, and A_3, respectively, are equal to the characteristic polynomials, i.e., $m_1(t) = (t-2)^2$, $m_2(t) = (t-2)(t-7)$, $m_3(t) = t-7$. But $m(t)$ is equal to the least common multiple of $m_1(t)$, $m_2(t)$, $m_3(t)$. Thus $m(t) = (t-2)^2(t-7)$.

412. B is block diagonal with blocks $B_1 = \begin{bmatrix} 3 & 1 \\ 0 & 3 \end{bmatrix}$ and $B_2 = \begin{bmatrix} 3 & 1 & 0 \\ 0 & 3 & 1 \\ 0 & 0 & 3 \end{bmatrix}$. Since B is triangular, $\Delta(t) = (t-3)^5$, which is the product of the characteristic polynomials of the blocks.

413. By Remark 10.6, the minimal polynomials of the blocks are $f(t) = (t-3)^2$ and $g(t) = (t-3)^3$. Then $m(t) = \gcd(f(t), g(t)) = (t-3)^3$.

414. Since C is triangular, $\Delta(t) = (t-\lambda)^5$.

415. Since $C - \lambda I = 0$, $m(t) = t - \lambda$.

416. Since T is invertible, it is also nonsingular; hence $\lambda \neq 0$. By the definition of an eigenvalue, there exists a nonzero vector $\mathbf{v}$ for which $T(\mathbf{v}) = \lambda\mathbf{v}$. Applying T^{-1} to both sides, we obtain $\mathbf{v} = T^{-1}(\lambda\mathbf{v}) = \lambda T^{-1}(\mathbf{v})$. Hence $T^{-1}(\mathbf{v}) = \lambda^{-1}\mathbf{v}$; that is, λ^{-1} is an eigenvalue of T^{-1}.

417. Suppose $S(\mathbf{v}) = \lambda_1(\mathbf{v})$ and $T(\mathbf{v}) = \lambda_2(\mathbf{v})$. Then $(S + T)(\mathbf{v}) = S(\mathbf{v}) + T(\mathbf{v}) = \lambda_1\mathbf{v} + \lambda_2\mathbf{v} = (\lambda_1 + \lambda_2)\mathbf{v}$. Thus $\mathbf{v}$ is an eigenvector of $S + T$ belonging to the eigenvalue $\lambda_1 + \lambda_2$.

418. Suppose $T(\mathbf{v}) = \lambda\mathbf{v}$. Then $(kT)(\mathbf{v}) = kT(\mathbf{v}) = k(\lambda\mathbf{v}) = (k\lambda)\mathbf{v}$. Thus $\mathbf{v}$ is an eigenvector of kT belonging to the eigenvalue $k\lambda$.

419. Since λ is an eigenvalue of T, there exists a nonzero vector $\mathbf{v}$ such that $T(\mathbf{v}) = \lambda\mathbf{v}$.

  (1) We have $T^2(\mathbf{v}) = T(T(\mathbf{v})) = T(\lambda\mathbf{v}) = \lambda(T(\mathbf{v})) = \lambda(\lambda\mathbf{v}) = \lambda^2\mathbf{v}$. Thus λ^2 is an eigenvalue of T^2.

  (2) Suppose $n > 1$ and the result holds for $n - 1$. Then $T^n(\mathbf{v}) = T(T^{n-1}(\mathbf{v})) = T(\lambda^{n-1}\mathbf{v}) = \lambda^{n-1}(T(\mathbf{v})) = \lambda^{n-1}(\lambda\mathbf{v}) = \lambda^n\mathbf{v}$. Thus λ^n is an eigenvalue of T^n.

420. There exists a nonzero vector $\mathbf{v}$ such that $T(\mathbf{v}) = \lambda\mathbf{v}$. Suppose $f(t) = a_n t^n + \cdots + a_1 t + a_0$. Then

$$f(T)(\mathbf{v}) = (a_n T^n + \cdots + a_1 T + a_0 I)(\mathbf{v}) = a_n T^n(\mathbf{v}) + \cdots + a_1 T(\mathbf{v}) + a_0 I(\mathbf{v})$$
$$= a_n \lambda^n \mathbf{v} + \cdots + a_1 \lambda\mathbf{v} + a_0\mathbf{v} = (a_n \lambda^n + \cdots + a_1 \lambda + a_0)(\mathbf{v})$$
$$= f(\lambda)\mathbf{v}.$$

Thus $f(\lambda)$ is an eigenvalue of $f(T)$.

Chapter 11: Canonical Forms

421. Each vector $\mathbf{w} = (a, b, 0)$ in the xy-plane W remains in W under the mapping W. Thus W is invariant under T. The restriction of T to W rotates each vector in W about the origin O.

422. A nonzero vector $\mathbf{w}' = (0, b, c)$ in W' does not remain in W' under T (unless $\theta = \pi$ or a multiple of π). Thus W' is not T-invariant.

423. For any vector $\mathbf{u} = (0, 0, z)$ in U, we have $T(\mathbf{u}) = \mathbf{u}$. Thus U is invariant under T. In fact, the restriction of T to U is the identity mapping on U.

424. A nonzero vector $\mathbf{u}' = (a, 0, 0)$ in U' does not remain in U' under T (unless $\theta = \pi$ or a multiple of π). Thus U' is not invariant under T.

425. If $\mathbf{v}$ is any nonzero eigenvector of T, then span$(\mathbf{v})$ is a one-dimensional invariant subspace of T. Conversely, if W is a one-dimensional invariant subspace of T, then any nonzero vector in W is an eigenvector of T.

426. Let $\mathbf{u} \in \text{Ker}(T)$. Then $T(\mathbf{u}) = \mathbf{0} \in \text{Ker}(T)$ since the kernel of T is a subspace of V. Thus $\text{Ker}(T)$ is invariant under T.

427. Since $T(\mathbf{v}) \in \text{Im}(T)$ for every $\mathbf{v} \in V$, it is certainly true if $\mathbf{v} \in \text{Im}(T)$. Hence the image of T is invariant under T.

428. Here $\Delta(t) = t^2 - 3t - 10 = (t - 5)(t + 2)$. There are two eigenvalues, $\lambda_1 = 5$ and $\lambda_2 = -2$. Setting $A - 5I = 0$ yields the eigenvector $\mathbf{v}_1 = (2, 1)$, and setting $A + 2I = 0$ yields the eigenvector $\mathbf{v}_2 = (1, -3)$. Thus the only invariant subspaces of A are $\{0\}$, $\mathbf{R}^2$, span$(2, 1)$ and span$(1, -3)$.

429. Here $\Delta(t) = t^2 + 16$ is the characteristic polynomial of A. There are no eigenvalues (in $\mathbf{R}$) and hence there are no eigenvectors. Thus there are no one-dimensional invariant subspaces. Accordingly, $\{0\}$ and $\mathbf{R}^2$ are the only A-invariant subspaces.

430. Since $\Delta(t) = t^2 + 16 = (t+4i)(t-4i)$, there are two eigenvalues, $\lambda_1 = 4i$ and $\lambda_2 = -4i$. Setting $\lambda_1 I - A = 0$ yields a nonzero solution $\mathbf{v}_1 = (2, 1-2i)$, and setting $\lambda_2 I - A = 0$ yields a nonzero solution $\mathbf{v}_2 = (2, 1+2i)$. Thus the only invariant subspaces are the following: $\{0\}$, $\mathbf{C}^2$, $W_1 = \text{span}(2, 1-2i)$, $W_2 = \text{span}(2, 1+2i)$.

431. $\mathbf{R}^3 = U + W$, since every vector in $\mathbf{R}^3$ is the sum of a vector in U and a vector in W. However, $\mathbf{R}^3$ is not the direct sum of U and W, since such sums are not unique, e.g., $(1, 2, 3) = (1, 1, 0) + (0, 1, 1) = (1, 3, 0) + (0, -1, 3)$.

432. Any vector $(a, b, c) \in \mathbf{R}^3$ can be written as the sum of a vector in U and a vector in Z in only one way, $(a, b, c) = (a, b, 0) + (0, 0, c)$. Thus $\mathbf{R}^3 = U \oplus Z$.

433. Any vector $(a, b, c) \in \mathbf{R}^3$ can be written as the sum of a vector in U and a vector in L in only one way, $(a, b, c) = (a-c, b-c, 0) + (c, c, c)$. Thus $\mathbf{R}^3 = U \oplus L$.

434. Any vector $(a, b, c) \in \mathbf{R}^3$ can be written as the sum of a vector in W and a vector in L in only one way, $(a, b, c) = (0, b-a, c-a) + (a, a, a)$. Thus $\mathbf{R}^3 = W \oplus L$.

435. Since $A^k = 0$ but $A^{k-1} \neq 0$, we have $m(t) = t^k$.

436. Since $m(t) = t^k$ is the minimum polynomial of A, only 0 is an eigenvalue of A.

437. Since the degree of the characteristic polynomial $\Delta(t)$ of A is n, $k = \deg(m(t)) \leq \deg(\Delta(t)) = n$. Thus $k \leq n$.

438. Since $A^k = 0$, we have A^k is singular. Recall that the product of nonsingular matrices is nonsingular; hence A must be singular.

439. Compute $A^2 = \begin{bmatrix} -1 & 0 & 1 \\ -1 & 0 & 1 \\ -1 & 0 & 1 \end{bmatrix}$ and $A^3 = 0$. Thus A is nilpotent of index 3.

440. Compute $B^2 = \begin{bmatrix} 2 & 6 & -4 \\ 2 & 6 & -4 \\ 2 & 6 & -4 \end{bmatrix}$ and $B^3 = \begin{bmatrix} 4 & 12 & -8 \\ 4 & 12 & -8 \\ 4 & 12 & -8 \end{bmatrix}$. Thus B is not nilpotent. (We do not need to test higher than the order of B.)

441. Compute $C^2 = 0$. Thus C is nilpotent of index 2.

442. Since the index of A is 3, its canonical form is $\begin{bmatrix} 0 & 1 & 0 \\ 0 & 0 & 1 \\ 0 & 0 & 0 \end{bmatrix}$.

443. B is not nilpotent, so it is not similar to any canonical nilpotent matrix.

444. Since the index of C is 2, its canonical form is $\operatorname{diag}\left(\begin{bmatrix} 0 & 1 \\ 0 & 1 \end{bmatrix},\ [0]\right)$.

445. Both $\Delta(t)$ and $m(t)$ are equal to $(t-7)^4$; that is, $\Delta(t) = m(t) = (t-7)^4$. Thus $\lambda = 7$ is the only eigenvalue.

446. Subtracting $\lambda = 7$ down the diagonal of A yields the corresponding system

$$-y = 0,\ -z = 0,\ -t = 0,\ 0 = 0.$$

There is only one free variable x; hence $\mathbf{v} = (1, 0, 0, 0)$ forms a basis of the eigenspace of $\lambda = 7$.

447. Here $\Delta(t) = (t+3)^3(t-5)^4$. The exponent 3 comes from the fact that there are three -3's on the diagonal, and the exponent 4 comes from the fact that there are four 5's on the diagonal. In particular, $\lambda_1 = -3$ and $\lambda_2 = 5$ are the eigenvalues.

448. Here $m(t) = (t+3)^3(t-5)^2$. The exponent 3 comes from the fact that 3 is the order of the largest block belonging to $\lambda_1 = -3$, and the exponent 2 comes from the fact that 2 is the order of the largest block belonging to $\lambda_2 = 5$. [Alternatively, $m(t)$ is the least common multiple of the minimal polynomials of the blocks.]

449. Each block contributes one eigenvector to S. Three such eigenvectors are

$$\mathbf{v}_1 = (1, 0, 0, 0, 0, 0, 0),\ \mathbf{v}_2 = (0, 0, 0, 1, 0, 0, 0),\ \mathbf{v}_3 = (0, 0, 0, 0, 0, 1, 0),$$

which correspond to the first, second, and third blocks, respectively. The entry 1 in each vector is the position of the first entry in the corresponding block.

450. Here $\Delta(t) = (t-4)^5(t-2)^3$, since there are five 4s on the diagonal and three 2s on the diagonal. Thus $\lambda_1 = 4$ and $\lambda_2 = 2$ are the eigenvalues of A.

451. Here $m(t) = (t-4)^3(t-2)^2$, since 3 is the order of the largest block in A belonging to $\lambda_1 = 4$ and 2 is the order of the largest block in A belonging to $\lambda_2 = 2$.

452. Here $d_1 = 2$ since there are two blocks belonging to $\lambda_1 = 4$. Also, $\mathbf{v}_1 = (1, 0, 0, 0, 0, 0, 0, 0)$ and $\mathbf{v}_2 = (0, 0, 0, 1, 0, 0, 0, 0)$ form a basis of E_1.

453. There are two blocks in A belonging to $\lambda_2 = 2$; hence $d_2 = 2$. Also, $\mathbf{w}_1 = (0, 0, 0, 0, 0, 1, 0, 0)$ and $\mathbf{w}_2 = (0, 0, 0, 0, 0, 0, 0, 1)$ form a basis of E_2.

454. Note that 2 is the order of the largest block in B belonging to $\lambda_1 = 4$ and 3 is the order of the largest block in B belonging to $\lambda_2 = 2$; hence $m(t) = (t-4)^2(t-2)^3$.

455. Although A and B have the same characteristic polynomial and the same eigenvalues, A and B are not equivalent since the diagonal blocks are different.

456. Note that 2 is the order of the largest block in B belonging to $\lambda_1 = 4$ and 3 is the order of the largest block in B belonging to $\lambda_2 = 2$; hence $m(t) = (t-4)^2(t-2)^3$.

457. There are three blocks in B belonging to $\lambda_1 = 4$; hence $d_1 = 3$. Also $\mathbf{v}_1 = (1, 0, 0, 0, 0, 0, 0, 0)$, $\mathbf{v}_2 = (0, 0, 1, 0, 0, 0, 0, 0)$, $\mathbf{v}_3 = (0, 0, 0, 0, 1, 0, 0, 0)$ form a basis of E_1.

458. There is only one block in B belonging to $\lambda_2 = 2$; hence $d_2 = 1$. Also, $\mathbf{w} = (0, 0, 0, 0, 0, 1, 0, 0)$ forms a basis of E_2.

459. Since $\Delta(t) = (t - 2)^4(t - 3)^3$, there must be four 2s on the diagonal and three 3s on the diagonal. Also, since $m(t) = (t - 2)^2(t - 3)^2$, there must be a block of order 2, and none larger, belonging to the eigenvalue 2; and there must be a block of order 2, and none larger, belonging to the eigenvalue 3. There are two possibilities, which follow:

$$\text{diag}\left(\begin{bmatrix} 2 & 1 \\ 0 & 2 \end{bmatrix}, \begin{bmatrix} 2 & 1 \\ 0 & 2 \end{bmatrix}, \begin{bmatrix} 3 & 1 \\ 0 & 3 \end{bmatrix}, [3]\right) \quad \text{or} \quad \text{diag}\left(\begin{bmatrix} 2 & 1 \\ 0 & 2 \end{bmatrix}, [2], [2], \begin{bmatrix} 3 & 1 \\ 0 & 3 \end{bmatrix}, [3]\right).$$

The first matrix occurs if T has two independent eigenvectors belonging to its eigenvalue 2, and the second matrix occurs if T has three independent eigenvectors belonging to 2.

460. Since $\Delta(t) = (t - 7)^5$ has degree 5, the matrix must have order 5 and have five 7s on the diagonal. Also, since $m(t) = (t - 7)^2$, there must be a block of order 2, and none higher. There are two possibilities, which follow:

$$\text{diag}\left(\begin{bmatrix} 7 & 1 \\ 0 & 7 \end{bmatrix}, \begin{bmatrix} 7 & 1 \\ 0 & 7 \end{bmatrix}, [7]\right) \quad \text{or} \quad \text{diag}\left(\begin{bmatrix} 7 & 1 \\ 0 & 7 \end{bmatrix}, [7], [7], [7]\right).$$

The first occurs if $\lambda = 7$ has geometric multiplicity 3, and the second occurs if $\lambda = 7$ has geometric multiplicity 4.

Chapter 12: Linear Operators on Inner Product Spaces

461. For every $\mathbf{u}, \mathbf{v} \in \mathbf{R}^n$, $\langle A\mathbf{u}, \mathbf{v} \rangle = (A\mathbf{u})^T \mathbf{v} = \mathbf{u}^T A^T \mathbf{v} = \langle \mathbf{u}, A^T \mathbf{v} \rangle$ Thus A^T is the adjoint of A.

462. For every $\mathbf{u}, \mathbf{v} \in \mathbf{C}^n$, $\langle B\mathbf{u}, \mathbf{v} \rangle = (B\mathbf{u})^T \mathbf{v} = \mathbf{u}^T B^T \mathbf{v} = \mathbf{u}^T \overline{\overline{B}^T} \overline{\mathbf{v}} = \mathbf{u}^T \overline{B^* \mathbf{v}} = \langle \mathbf{u}, B^* \mathbf{v} \rangle$. Thus B^* is the adjoint of B.

463. Take the conjugate transpose of A to get $A^* = \begin{bmatrix} 2 - 3i & 6 + 9i \\ 5 + 4i & 2 - 7i \end{bmatrix}$.

464. The conjugate transpose gives us $B^* = \begin{bmatrix} 3 - 7i & 18 & 4 + i \\ -7i & 6 - i & 2 - 3i \\ 8 + i & 7 + 9i & 6 + 3i \end{bmatrix}$.

465. Since C is real, the adjoint C^* is simply the transpose of C. Thus $C^* = C^T = \begin{bmatrix} 1 & 4 & 7 \\ 2 & 5 & 7 \\ 3 & 6 & 7 \end{bmatrix}$.

466. For any $\mathbf{u}, \mathbf{v} \in V, \langle (S+T)(\mathbf{u}), \mathbf{v} \rangle = \langle S(\mathbf{u}) + T(\mathbf{u}), \mathbf{v} \rangle = \langle S(\mathbf{u}), \mathbf{v} \rangle + \langle T(\mathbf{u}), \mathbf{v} \rangle = \langle \mathbf{u}, S^*(\mathbf{v}) \rangle +$ $\langle \mathbf{u}, T^*(\mathbf{v}) \rangle = \langle \mathbf{u}, S^*(\mathbf{v}) + T^*(\mathbf{v}) \rangle = \langle \mathbf{u}, (S^* + T^*)(\mathbf{v}) \rangle$. The uniqueness of the adjoint implies $(S + T)^* = S^* + T^*$.

467. For any $\mathbf{u}, \mathbf{v} \in V, \langle (kT)(\mathbf{u}), \mathbf{v} \rangle = \langle kT(\mathbf{u}), \mathbf{v} \rangle = k \langle T(\mathbf{u}), \mathbf{v} \rangle = k \langle \mathbf{u}, T^*(\mathbf{v}) \rangle = \langle \mathbf{u}, \bar{k}T^*(\mathbf{v}) \rangle =$ $\langle \mathbf{u}, (\bar{k}T^*)(\mathbf{v}) \rangle$. The uniqueness of the adjoint implies $(kT)^* = \bar{k}T^*$.

468. For every $\mathbf{u}, \mathbf{v} \in V, \langle (ST)(\mathbf{u}), \mathbf{v} \rangle = \langle S(T(\mathbf{u})), \mathbf{v} \rangle = \langle T(\mathbf{u}), S^*(\mathbf{v}) \rangle = \langle \mathbf{u}, T^*(S^*(\mathbf{v})) \rangle =$ $\langle \mathbf{u}, (T^*S^*)(\mathbf{v}) \rangle$. The uniqueness of the adjoint implies $(ST)^* = T^*S^*$.

469. For every $\mathbf{u}, \mathbf{v} \in V, \langle T^*(\mathbf{u}), \mathbf{v} \rangle = \overline{\langle \mathbf{v}, T^*(\mathbf{u}) \rangle} = \overline{\langle T(\mathbf{v}), \mathbf{u} \rangle} = \langle \mathbf{u}, T(\mathbf{v}) \rangle$. The uniqueness of the adjoint implies $(T^*)^* = T$.

470. Let $\mathbf{u} \in W^\perp$. If $\mathbf{w} \in W$, then $T(\mathbf{w}) \in W$ and so $\langle \mathbf{w}, T^*(\mathbf{u}) \rangle = \langle T(\mathbf{w}), \mathbf{u} \rangle = 0$. Thus $T^*(\mathbf{u}) \in W^\perp$ since it is orthogonal to every $\mathbf{w} \in W$. Hence $W^\perp$ is invariant under T^*.

471. (A) For every $\mathbf{u}, \mathbf{v} \in V, \langle I(\mathbf{u}), \mathbf{v} \rangle = \langle \mathbf{u}, \mathbf{v} \rangle = \langle \mathbf{u}, I(\mathbf{v}) \rangle$; hence $I^* = I$.
(B) For every $\mathbf{u}, \mathbf{v} \in V, \langle 0(\mathbf{u}), \mathbf{v} \rangle = \langle 0, \mathbf{v} \rangle = 0 = \langle \mathbf{u}, 0 \rangle = \langle \mathbf{u}, 0(\mathbf{v}) \rangle$; hence $0^* = 0$.

472. $I = I^* = (TT^{-1})^* = (T^{-1})^*T^*$; hence $(T^{-1})^* = (T^*)^{-1}$.

473. Let $\mathbf{v}$ be a nonzero eigenvector of T belonging to λ, that is, $T(\mathbf{v}) = \lambda \mathbf{v}$ with $\mathbf{v} \neq 0$; hence $\langle \mathbf{v}, \mathbf{v} \rangle$ is positive. We show that $\lambda \langle \mathbf{v}, \mathbf{v} \rangle = \bar{\lambda} \langle \mathbf{v}, \mathbf{v} \rangle$:

$$\lambda \langle \mathbf{v}, \mathbf{v} \rangle = \langle \lambda \mathbf{v}, \mathbf{v} \rangle = \langle T(\mathbf{v}), \mathbf{v} \rangle = \langle \mathbf{v}, T^*(\mathbf{v}) \rangle = \langle \mathbf{v}, T(\mathbf{v}) \rangle = \langle \mathbf{v}, \lambda \mathbf{v} \rangle = \bar{\lambda} \langle \mathbf{v}, \mathbf{v} \rangle.$$

But $\langle \mathbf{v}, \mathbf{v} \rangle \neq 0$; hence $\lambda = \bar{\lambda}$ and so λ is real.

474. Suppose $T(\mathbf{v}) = \lambda \mathbf{v}$ and $T(\mathbf{w}) = \mu \mathbf{w}$ where $\lambda \neq \mu$. We show that $\lambda \langle \mathbf{v}, \mathbf{w} \rangle = \mu \langle \mathbf{v}, \mathbf{w} \rangle$:

$$\lambda \langle \mathbf{v}, \mathbf{w} \rangle = \langle \lambda, \mathbf{w} \rangle = \langle T(\mathbf{v}), \mathbf{w} \rangle = \langle \mathbf{v}, T(\mathbf{w}) \rangle = \langle \mathbf{v}, \mu \mathbf{w} \rangle = \bar{\mu} \langle \mathbf{v}, \mathbf{w} \rangle = \mu \langle \mathbf{v}, \mathbf{w} \rangle.$$

(The last step uses the fact that μ is real, so $\bar{\mu} = \mu$.) But $\lambda \neq \mu$; hence $\langle \mathbf{u}, \mathbf{w} \rangle = 0$, as claimed.

475. Here $\Delta(t) = t^2 - \text{tr}(A) + |A| = t^2 - 4t - 21 = (t - 7)(t + 3)$, Thus $\lambda = 7$ and $\lambda = -3$ are the eigenvalues of A.
  (1) Subtract $\lambda = 7$ down the diagonal of A to obtain the corresponding homogeneous system $\begin{matrix} -2x + 4y = 0 \\ 4x - 8y = 0 \end{matrix}$ or, simply, $x - 2y = 0$. A nonzero solution is $\mathbf{u}_1 = (2, 1)$.
  (2) Subtract $\lambda = -3$ (or add 3) down the diagonal of A to obtain the corresponding homogeneous system $\begin{matrix} 8x + 4y = 0 \\ 4x + 2y = 0 \end{matrix}$ or, simply, $2x + y = 0$. A nonzero solution is $\mathbf{u}_2 = (1, -2)$.
Thus $S = \{\mathbf{u}_1 = (2, 1), \mathbf{u}_2 = (1, -2).\}$ (As expected, u_1 and u_2 are orthogonal.)

476. Normalize u_1 and u_2 to obtain the unit vectors $\hat{u}_1 = (2/\sqrt{5}, 1/\sqrt{5})$ and $\hat{u}_2 = (1/\sqrt{5}, -2/\sqrt{5})$.

Let P be the matrix whose columns are $\hat{u}_1$ and $\hat{u}_2$. Then

$$P = \begin{bmatrix} 2/\sqrt{5} & 1/\sqrt{5} \\ 1/\sqrt{5} & -2/\sqrt{5} \end{bmatrix} \quad \text{and} \quad D = P^{-1}AP = \begin{bmatrix} 7 & 0 \\ 0 & -3 \end{bmatrix}.$$

477. Here $\text{tr}(C) = 2 + 5 + 2 = 9$, $C_{11} = \begin{vmatrix} 5 & 2 \\ 2 & 2 \end{vmatrix} = 6$, $C_{22} = \begin{vmatrix} 2 & 1 \\ 1 & 2 \end{vmatrix} = 3$, $C_{33} = \begin{vmatrix} 2 & 2 \\ 2 & 5 \end{vmatrix} = 6$.

Hence $C_{11} + C_{22} + C_{33} = 15$, and $|C| = 20 + 4 + 4 - 5 - 8 - 8 = 7$. Thus $\Delta(t) = t^3 - 9t^2 + 15t - 7$. If $\Delta(t)$ has a rational root it must divide the constant 7, so it must be ± 1 or ± 7. Testing, we get 1 as a root. So $\Delta(t) = (t-1)(t^2 - 8t + 7) = (t-1)^2(t-7)$.

Thus the eigenvalues of C are $\lambda = 1$ (with multiplicity two) and $\lambda = 7$ (with multiplicity one).

478. (1) Subtract $\lambda = 1$ down the diagonal of C to obtain the homogeneous system $x + 2y + z = 0$, $2x + 4y + 2z = 0$, $x + 2y + z = 0$. That is, $x + 2y + z = 0$. One solution is $v_1 = (0, 1, -2)$. We seek a second solution $v_2 = (a, b, c)$ which is orthogonal to v_1. This means $a + 2b + c = 0$ and also $b - 2c = 0$. One such solution is $v_2 = (5, -2, -1)$.

(2) Subtract $\lambda = 7$ down the diagonal of C to obtain the homogeneous system $-5x + 2y + z = 0$, $2x - 2y + 2z = 0$, $x + 2y - 5z = 0$. The system yields a nonzero solution $v_3 = (1, 2, 1)$. (As expected, the eigenvector v_3 is orthogonal to v_1 and v_2.) Thus $S = \{v_1 = (0, 1, -2), v_2 = (5, -2, -1), v_3 = (1, 2, 1)\}$. (We emphasize that S is not unique.)

479. Normalize the vectors v_1, v_2, v_3 to obtain the unit solution $\hat{v}_1 = (0, 1/\sqrt{5}, -2/\sqrt{5})$, $\hat{v}_2 = (5/\sqrt{30}, -2/\sqrt{30}, -1/\sqrt{30})$, $\hat{v}_3 = (1/\sqrt{6}, 2/\sqrt{6}, 1/\sqrt{6})$. Let P be the matrix whose columns are $\hat{v}_1, \hat{v}_2, \hat{v}_3$. Then $D = P^{-1}CP = \text{diag}(1, 1, 7)$.

480. Since C is the symmetric matrix that represents q, use the above matrix P to obtain the following change of coordinates: $x = 5/\sqrt{30}\, s + 1/\sqrt{6}\, t$, $y = 1/\sqrt{5}\, r - 2/\sqrt{30}\, s + 2/\sqrt{6}\, t$, $z = -2/\sqrt{5}\, r - 1/\sqrt{30}\, s + 1/\sqrt{6}\, t$. Under this change of coordinates, q is transformed into the diagonal form $q(r, s, t) = r^2 + s^2 + 7t^2$.

481. For any $v \in V$, we have $\|T(v)\|^2 = \langle T(v), T(v) \rangle = \langle v, T^2(v) \rangle = \langle v, 0v \rangle = \langle v, 0 \rangle = 0$. Thus $\|T(v)\| = 0$ and hence $T(v) = 0$. Since $T(v) = 0$ for every $v \in V$, we have $T = 0$.

482. (A) $(T^*T)^* = T^*T^{**} = T^*T$, and hence T^*T is self-adjoint. Also, $(TT^*)^* = T^{**}T^* = TT^*$, and hence TT^* is self-adjoint.

(B) $(T + T^*)^* = T^* + T^{**} = T^* + T = T + T^*$; hence $T + T^*$ is self-adjoint.

483. As pictured in Figure 12.1(a), the length (distance from the origin) of v is preserved under the rotation T. Thus T is an orthogonal operator.

484. As pictured in Figure 12.1(b), the length of **v** is preserved under the reflection F. Thus F is orthogonal.

485. Let **v** be a nonzero eigenvector of T belonging to λ, that is, $T(\mathbf{v}) = \lambda\mathbf{v}$ with $\mathbf{v} \neq \mathbf{0}$; hence $\langle \mathbf{v}, \mathbf{v} \rangle$ is positive. We show that $\lambda\bar{\lambda}\langle \mathbf{v}, \mathbf{v} \rangle = \langle \mathbf{v}, \mathbf{v} \rangle$: $\lambda\bar{\lambda}\langle \mathbf{v}, \mathbf{v} \rangle = \langle \lambda\mathbf{v}, \lambda\mathbf{v} \rangle = \langle T(\mathbf{v}), T(\mathbf{v}) \rangle = \langle \mathbf{v}, T^*T(\mathbf{v}) \rangle = \langle \mathbf{v}, I(\mathbf{v}) \rangle = \langle \mathbf{v}, \mathbf{v} \rangle$. But $\langle \mathbf{v}, \mathbf{v} \rangle \neq 0$; hence $\lambda\bar{\lambda} = 1$ and so $|\lambda| = 1$.

486. Suppose $A = \begin{bmatrix} a & b \\ c & d \end{bmatrix}$. Since A is orthogonal, its rows form an orthonormal set; hence $a^2 + b^2 = 1$, $c^2 + d^2 = 1$, $ac + bd = 0$, $ad - bc = 1$. The last equation follows from $\det(A) = 1$. We consider separately the cases $a = 0$ and $a \neq 0$.
If $a = 0$, the first equation gives $b^2 = 1$ and therefore $b = \pm 1$. Then the fourth equation gives $c = -b = \pm 1$, and the second equation yields $1 + d^2 = 1$, or $d = 0$. Thus

$$A = \begin{bmatrix} 0 & 1 \\ -1 & 0 \end{bmatrix} \quad \text{or} \quad \begin{bmatrix} 0 & -1 \\ 1 & 0 \end{bmatrix}.$$

The first alternative has the required form with $\theta = -\pi/2$, and the second alternative has the required form with $\theta = \pi/2$.
If $a \neq 0$, the third equation can be solved to give $c = -bd/a$. Substituting this into the second equation gives $b^2d^2/a^2 + d^2 = 1$, or $b^2d^2 + a^2d^2 = a^2$, or $(b^2 + a^2)d^2 = a^2$, or $a^2 = d^2$; therefore $a = d$ or $a = -d$. If $a = -d$, then the third equation yields $c = b$ and so the fourth equation gives $-a^2 - c^2 = 1$, which is impossible. Thus $a = d$. But then the third equation gives $b = -c$, and so

$$A = \begin{bmatrix} a & -c \\ c & a \end{bmatrix}.$$

Since $a^2 + c^2 = 1$, there is a real number θ such that $a = \cos\theta$, $c = \sin\theta$, and hence A has the required form in this case also.

487. By definition, $P = S^*S$ for some operator S. Hence $P^* = (S^*S)^* = S^*S^{**} = S^*S = P$. Thus P is self-adjoint.

488. (A) Since $|A| = 0$, A is not positive definite. However, A is positive, since $a = 1$, $d = 1$, and $|A| = 0$ are nonnegative.
    (B) Since $a = 3$, $d = 3$, and $|B| = 8$ are positive, B is positive definite (and hence positive).
    (C) Since C is not self-adjoint (i.e., $C^T \neq C$), C is neither positive definite nor positive.

489. (A) Since $a = 2$, $d = 2$, and $|D| = 3$ are positive, D is positive definite (and hence positive).
    (B) Since $|E| = 0$, E is not positive definite. However, E is positive, since $a = 1$, $d = 1$, and $|E| = 0$ are nonnegative.
    (C) Since $|F| = -3$, F is neither positive definite nor positive.

490. (A) Since T is positive, T is self-adjoint; hence λ is real. Let $\mathbf{v}$ be a nonzero eigenvector of T belonging to λ [that is, $T(\mathbf{v}) = \lambda\mathbf{v}$ with $\mathbf{v} \neq 0$]; hence $\langle \mathbf{v}, \mathbf{v} \rangle$ is positive. Since T is positive, $T = S^*S$ for some operator S. We show that $\lambda\langle \mathbf{v}, \mathbf{v} \rangle = \langle S(\mathbf{v}), S(\mathbf{v}) \rangle$: $\lambda\langle \mathbf{v}, \mathbf{v} \rangle = \langle \lambda\mathbf{v}, \mathbf{v} \rangle = \langle T(\mathbf{v}), \mathbf{v} \rangle = \langle S^*S(\mathbf{v}), \mathbf{v} \rangle = \langle S(\mathbf{v}), S(\mathbf{v}) \rangle$. Since $\langle S(\mathbf{v}), S(\mathbf{v}) \rangle$ is nonnegative and $\langle \mathbf{v}, \mathbf{v} \rangle$ is positive, we have that λ is nonnegative, as required.

(B) Since T is positive definite, T is self-adjoint; hence λ is real. Let $\mathbf{v}$ be a nonzero eigenvector of T belonging to λ,[that is, $T(\mathbf{v}) = \lambda\mathbf{v}$ with $\mathbf{v} \neq 0$]; hence $\langle \mathbf{v}, \mathbf{v} \rangle$ is positive. Since T is positive definite, $T = S^*S$ for some nonsingular operator S. Thus $S(\mathbf{v}) \neq \mathbf{0}$, and hence $\langle S(\mathbf{v}), S(\mathbf{v}) \rangle$ is positive. We show that $\lambda\langle \mathbf{v}, \mathbf{v} \rangle = \langle S(\mathbf{v}), S(\mathbf{v}) \rangle$: $\lambda \langle \mathbf{v}, \mathbf{v} \rangle = \langle \lambda\mathbf{v}, \mathbf{v} \rangle = \langle T(\mathbf{v}), \mathbf{v} \rangle = \langle S^*S(\mathbf{v}), \mathbf{v} \rangle = \langle S(\mathbf{v}), S(\mathbf{v}) \rangle$.

Since $\langle \mathbf{v}, \mathbf{v} \rangle$ and $\langle S(\mathbf{v}), S(\mathbf{v}) \rangle$ are both positive, we have that λ is positive, as required.

491. Since $a_{11} = 5$, $a_{22} = 5$, and $|A| = 24$ are positive, A is a positive definite matrix.

492. Here $\Delta(t) = t^2 - \text{tr}(A) + |A| = t^2 - 10t + 24 = (t - 6)(t - 4)$. Thus $\lambda = 6$ and $\lambda = 4$ are the eigenvalues of A.

(1) Subtract $\lambda = 6$ down the diagonal of A to obtain the corresponding homogeneous system $-x + y = 0$ and $x - y = 0$. A nonzero solution is $\mathbf{u}_1 = (1, 1)$. Normalize $\mathbf{u}_1$ to obtain the unit solution $\hat{\mathbf{u}}_1 = (1/\sqrt{2}, 1/\sqrt{2})$.

(2) Subtract $\lambda = 4$ down the diagonal of A to obtain the corresponding homogeneous system $x + y = 0$ and $x + y = 0$. A nonzero solution is $\mathbf{u}_2 = (1, -1)$. Normalize $\mathbf{u}_2$ to obtain the unit solution $\hat{\mathbf{u}}_2 = (1/\sqrt{2}, -1/\sqrt{2})$.

Let Q be the matrix whose columns are $\hat{\mathbf{u}}_1$ and $\hat{\mathbf{u}}_2$. Then

$$Q = \begin{bmatrix} 1/\sqrt{2} & 1/\sqrt{2} \\ 1/\sqrt{2} & -1/\sqrt{2} \end{bmatrix} \quad \text{and} \quad Q^{-1}AQ = \begin{bmatrix} 6 & 0 \\ 0 & 4 \end{bmatrix}.$$

493. Take the square root of the diagonal entries to get $S = \begin{bmatrix} \sqrt{6} & 0 \\ 0 & 2 \end{bmatrix}$.

494. We have $B = Q^T AQ = Q^{-1}AQ$; hence $A = QBQ^{-1} = QBQ^T$. Then $T^2 = (QSQ^T) \times (QSQ^T) = (QSQ^{-1})(QSQ^{-1}) = QS^2Q^{-1} = QBQ^{-1} = A$. Also, T is positive definite. Thus T is the positive square root of A.

495. Compute

$$AA^* = \begin{bmatrix} 1 & 1 \\ i & 3+2i \end{bmatrix}\begin{bmatrix} 1 & -i \\ 1 & 3-2i \end{bmatrix} = \begin{bmatrix} 2 & 3-3i \\ 3+3i & 14 \end{bmatrix}$$

$$A^*A = \begin{bmatrix} 1 & -i \\ 1 & 3-2i \end{bmatrix}\begin{bmatrix} 1 & 1 \\ i & 3+2i \end{bmatrix} = \begin{bmatrix} 2 & 3-3i \\ 3+3i & 14 \end{bmatrix}.$$

Since $AA^* = A^*A$, the matrix A is normal.

496. Compute

$$BB^* = \begin{bmatrix} 1 & i \\ 0 & 1 \end{bmatrix}\begin{bmatrix} 1 & 0 \\ -i & 1 \end{bmatrix} = \begin{bmatrix} 2 & i \\ -i & 1 \end{bmatrix} \qquad B^*B = \begin{bmatrix} 1 & 0 \\ -i & 1 \end{bmatrix}\begin{bmatrix} 1 & i \\ 0 & 1 \end{bmatrix} = \begin{bmatrix} 1 & i \\ -i & 2 \end{bmatrix}.$$

Since $BB^* \neq B^*B$, the matrix B is not normal.

497. Compute

$$CC^* = \begin{bmatrix} 1 & i \\ 1 & 2+i \end{bmatrix}\begin{bmatrix} 1 & 1 \\ -i & 2-i \end{bmatrix} = \begin{bmatrix} 2 & 2+2i \\ 2-2i & 6 \end{bmatrix}$$

$$C^*C = \begin{bmatrix} 1 & 1 \\ -i & 2-i \end{bmatrix}\begin{bmatrix} 1 & i \\ 1 & 2+i \end{bmatrix} = \begin{bmatrix} 2 & 2+2i \\ 2-2i & 6 \end{bmatrix}.$$

Since $CC^* = C^*C$, the matrix C is normal.

498. We show that $T - \lambda I$ commutes with its adjoint:

$$(T - \lambda I)(T - \lambda I)^* = (T - \lambda I)(T^* - \bar{\lambda} I) = TT^* - \lambda T^* - \bar{\lambda} T + \lambda \bar{\lambda} I$$
$$= T^*T - \bar{\lambda} T - \lambda T^* + \bar{\lambda} \lambda I = (T^* - \bar{\lambda} I)(T - \lambda I)$$
$$= (T - \lambda I)^*(T - \lambda I).$$

Thus $T - \lambda I$ is normal.

499. We show that $\lambda_1 \langle \mathbf{v}, \mathbf{w} \rangle = \lambda_2 \langle \mathbf{v}, \mathbf{w} \rangle$: $\lambda_1 \langle \mathbf{v}, \mathbf{w} \rangle = \langle \lambda_1 \mathbf{v}, \mathbf{w} \rangle = \langle T(\mathbf{v}), \mathbf{w} \rangle = \langle \mathbf{v}, T^*(\mathbf{w}) \rangle = \langle \mathbf{v}, \lambda_2 \mathbf{w} \rangle = \lambda_2 \langle \mathbf{v}, \mathbf{w} \rangle$. But $\lambda_1 \neq \lambda_2$; hence $\langle \mathbf{v}, \mathbf{w} \rangle = 0$.

500. Let $E_1 = \text{diag}(1, 0, 0, 0)$, $E_2 = \text{diag}(0, 1, 1, 0)$, $E_3 = \text{diag}(0, 0, 0, 1)$. Then
   (i) $A = 2E_1 + 3E_2 + 5E_3$,
   (ii) $E_1 + E_2 + E_3 = I$,
   (iii) $E_i^2 = E_i$, and
   (iv) $E_i E_j = 0$ for $i \neq j$.